Life is Balanced
LIVING IS NOT!

Life is Balanced
LIVING IS NOT!

Prof. G. Lakshman

PARTRIDGE
A Penguin Company

ISBN: Softcover 978-1-4828-1136-0
Ebook 978-1-4828-1135-3

Partridge books may be ordered through booksellers or by contacting:

Partridge India
Penguin Books India Pvt.Ltd
11, Community Centre, Panchsheel Park, New Delhi 110017
India
www.partridgepublishing.com
Phone: 000.800.10062.62

This book is targeted to touch a billion lives and bring about transformations in people and in society.

Prof G. Lakshman

FOREWORD

I am indeed privileged to write a 'FOREWORD' for Prof.Lakshman's book *"Life is balanced, Living is not"*. It is an excellent book and a guide for all who wish to live a balanced life with spiritual and physical success. Prof.Lakshman is a teacher and author. He receives ideas, thoughts and knowledge from everywhere analyzing and accepting those which add value to life in traditional Indian philosophy "Aano bhadra krtavo yantu vishwatah".

He studies, and works on the principle of analyze, experiment and experience and then translates into a teacher's language and style with single minded devotion to reach his experience and wisdoms to others.

"Life is balanced, Living is not" is one such book authored by him based on his own experiments and analyses and presents Common Denominators of Success and Leadership. He starts with a detailed definition of success and explains true meaning of success in professional and personal life. Life is divine. We are spiritual beings with Physical form. Everyone is blessed with life as a human being with a purpose, a dharma to perform. Identifying one's God given gifts, talents and acquired skill & knowledge

and putting these in the service of society and nature is the journey to success.

The book is divided into eight chapters, each chapter dealing with a process leading to success in life. Under each section, there are exercise, which may be carried out under the supervision of a teacher, so that the subject covered and process explained could be experienced by the participants. In this way, it is a different kind of book as it is not only presenting topics, but also guides the reader through well written exercises. I understood from Prof.Lakshman that these exercises were carried out by groups of students under his guidance and refined based on the experience gained through experiments. The book can therefore be used by a reader for self-teaching the various methods suggested by Prof.Lakshman. The book deals with success in a holistic manner and I am sure it will be a very useful book on, "**The Principles of Self Science"** for those who want to achieve success in a holistic way.

Dr C.G.Krishnadas Nair
Padmashree Awardee
Former Chairman, Hindustan Aeronautics LTD. (Hal)
President, Society of Indian Aerospace Technologies & Industries (SIATI)

FOREWORD

If any enterprise has to succeed in the true sense of the term, one has to pay equal attention to the deed as well as the doer.

The deed will be perfect if done as per the rules of the game.

The doer will be perfect if his intentions are clean and the techniques adopted are efficient.

Then and then alone, the greatest good of the greatest number will be achieved.

Hence we have to pay as much attention to the reformation or transformation of the individual as to the system itself.

This is what Prof Lakshman is trying to impress upon the readers in this work, Life is Balanced Living is not!

In eight short chapters with catchy titles, he has tried to amalgamate the ancient values of morality and ethics with the modern principles of management. He has prescribed several practical methods such as self enquiry, scribbling notes to remember things, self evaluation including one's behavioral pattern towards others. Above all, he advises the readers to search for and discover a meaning for one's life.

In a way, we can say that the spirit behind the famous saying of Swami Vivekananda viz, "for the liberation of oneself and for the service of mankind (one should strive)" has been reflected in this short treatise on successful Management.

Swami Harshananda,
Adhyaksha
Ramakrishna Mutt
Bangalore-560019

TABLE OF CONTENTS

From the Author's desk

This book serves as a self-help guide for personal growth. The eight chapters need not be read sequentially, but each chapter builds on the other. To keep the interest of the reader alive, each of the chapter has apt stories to convey the theme. The activity sheet in each chapter also provides the individual, insights through experiential learning.

The insights shared in each chapter is like clay, it can be mud in shoes, brick in building or a statue that will inspire all who see it. The clay is the same. The result is dependent on how it is used.

The Pause, Reflect and Scribble notes are touch points for the adult readers to delve deeper, and eventually your revelation will guide you to Connect to your true nature, for your healing.

The book can also be used as an additional reading material on the subject of ethics, values, wellness and integration of body, mind and soul. The book is apt course ware on, **The Principles of Self Science** for graduate and post graduate programme.

Prof G.Lakshman
www.glakshman.com
glakshman1@gmail.com

ACKNOWLEDGEMENTS

Over the years I have learned from, and been influenced by many individuals. I would like to acknowledge and give a public praising to the following people:

My parents—for inculcating the values of personal honesty and transparency

Dr Chenraj Jain for what he taught me about unlocking human potential.

My sincere appreciation to Dr Nagendra for what he taught me about skillfully weaving the various Yoga sciences into Self sciences.

I am also grateful to Ms. Kalpana Shashimohan for proofreading the text and making the necessary corrections.

I would like to acknowledge all the great leaders, thinkers, and sages of all religion who have walked this earth, and whose words and thoughts have been shared and quoted widely throughout the book.

Last, but certainly not least, I bow down to my Guru, Paramahamsa Yogananda for his guidance, to all humanity.

CHAPTER 1

The self science Model

Life is half spent before one knows what it is!

-French Proverb

A Brigadier once said, "I can command the whole brigade, but I cannot say a word to my wife and my children." The other day I heard the Vice Chancellor of a university commenting likewise, in a pensive mood, "Thousands of students flock to hear my lecture in my campus but the moment I open my mouth at home my children run away."

These people are successful in their profession, but in their domestic lives they have no sense of success. They are experts when it comes to handling some problems, but are incompetent when it comes to facing life as a whole—such indeed is the case with most of us

In the 21st century, the most popular books are books on Management and books on Self—Help. A lot of literature has also been written on how to be successful.

Management courses such as effective communication, business management teach people to be effective, productive and successful. But each of these courses deals with only one particular aspect of success.

We divide life into many small compartments and try to live in this piecemeal fashion. Merely accomplishing a task that we undertake cannot be the definition of success.

True meaning of Success:

Following the path of righteousness which is beneficial to the individual and the society, and attaining the end result with peace of mind and happiness is success. The joy of living, which is born of a sense of contentment in life, can be termed as success. **All living beings except humans know that the principle business of life is to enjoy day—to—day living**.

Success has to transform and enlighten the individual. Success is all about anchoring to peace and providing succor to near and dear ones and society at large. True meaning of success is the ability to overcome all kinds of problems-problems related to the family, health, wealth and relationships.

There is an ancient Indian story of two friends Mr. Vinod and Mr. Binod, who were poles apart in their behavior and attitude. Vinod was the "contemplative" type and Binod was the "activist" type.

One day, while Vinod was traveling in a boat with other passengers, they encountered windy and stormy weather. The boat pitched up and down and the terrified passengers were contemplating jumping out and swimming ashore. One by one every passenger including the boatman jumped for safety. But our friend Vinod was lazy to act and went into analysis and thought that everything would end well soon. He met his death when the boat crashed against a big rock. This happened because he was weak in energy and was overcome by idleness. Vinod would not act immediately but would move into 'mental gymnasium**. He never plunged from thought to action.**

The other friend Binod worked tirelessly and lived a very active life. He was of the opinion that spending time on simple joys of life like walking on the seashore early morning, watching the sunset, spending time with the family was a waste of time. He wanted to be productive so he was running from one activity to another all the time. Being in a hurry to accomplish more, he was always stressed. This hectic lifestyle took a toll on his health. One day when he developed fever and failed to give time to himself for healing sickness attacked him from inside and he also went to his deathbed. **He failed to slow down his pace of life.**

Nature shall not tolerate any imbalance for long.

Developing a successful personality and living life with a purpose is to strike the balance between the two aspects of our personality. Long phases of thinking, and analysis, sliding to reflection and contemplation, should translate to a growing need to move, to act on the insights. Inviting and involving others are also critical. Remember; when you are low and depressed, shake yourself by moving your body, exercise, jog—move into action and interaction with others because the seeds of ideas germinate better when shared with others. Then the intense build up of action should lead to conscious slow down—sliding to phases of thinking and reflection. This cyclic process of **thinking-action-reflection** has to be applied diligently in our life for progress and growth.

Success must bring enhancement of personality, purification of mind, joy and happiness in living.

Unfolding our potential is our true purpose:

Everyday life situations and experiences teach us a lot. But due to lack of impartial introspection and resisting change, sometimes one just slides from bad to worse conditions in health, wealth, relationship and personal confidence and happiness. There is actually no quick route to success. Remind yourself again and again that genius is 1% inspiration and 99% perspiration. Take courage from the fact that many people who later turned out to be great were quite ordinary in childhood but worked their way up. They chose and held on to choices which were turning points in their lives. **Remember they identified with their purpose at different stages of life and redefined their potential.**

Mahatma Gandhi made a huge impact during Indian Independence struggle in the 19th century. He was fondly known as, Bapuji or Father of the Nation. He was quite ordinary as a child and even as a young man. He went to England and studied law, but did not succeed in establishing his law practice in India. Luckily, he managed to get a one year assignment to serve as a lawyer's assistant in South Africa.

What changed his life entirely and started the transformation that was to take place later in his life was the reading of the book, "Unto this last" by John Ruskin. This is what Gandhi has to say about it in his autobiography: "The book was impossible to put aside, once I had begun it. It gripped me. Johannesburg to Durban was a 24 hour journey. The train reached there in the evening. I could not get any sleep that night. I was determined to change my life in accordance with the book. I believed that I discovered some of my deepest convictions reflected in this book of Ruskin, and that is why it so captured me and made me transform my life . . . . I arose with the dawn, ready to reduce these principles to practice."

Victor Hugo rightly says, "No army can withstand the strength of an idea whose time has come."

In this era of information overload what is important is quality time, quality thoughts and quest within oneself. **Your quest will drive you to your purpose and unfold your potential.**

I would like to quote what Lord Buddha says, "The enlightened can only show us the path. It is for you to walk on the path of experiential wisdom that each individual earns for himself."

This book contains some distilled essence of living principles that successful leaders have learned and practiced. One such principle is Recovery . . . the ability to bounce back; and moving from failure to success. In a day there are many hurt moments and failed transactions, but one has to recover from them quickly and move on by practicing present moment awareness. People out there in the world **"steal our goat"** (peace of mind, happiness & dreams.)The world tramples all over us and pushes us down the moment we ask why . . . . Or why not . . . . !

But yet one needs to unlock one's mind and scale new heights. One needs focus, and clarity in one's dreams. To help in this endeavor, one has to get inspiration and positive energy from likeminded friends and relatives and our personal learning journal.

Pause, Reflect, and Scribble

____
____
____
____
____
____
____
____
____
____

The Self Science Model

The book," Life is Balanced Living is not!" is an attempt in the direction of giving clarity to the purpose of living and transforming a billion lives. People are in constant stress to achieve social status and material prosperity. As a fall—out, human relationships, family bonding and value systems are affected. 21 st century lifestyles translate into lesser physical exercise, erratic sleep patterns, and anxiety disorders. Stress and work pressure eat into our time and we end up spending little quality time with our family, friends, and loved ones, which results in broken relationships and superficial interactions.

Lack of knowledge of Self science also results in poor interpersonal relationships and lack of compassion towards one another.

People find it difficult even to make basic sacrifices and adjustments in life. Problems are further compounded by falling prey to negative emotions—greed, anger, jealousy, hatred, lust, ... and so on.

The purpose of stating this is not to paint a grim picture but to help individuals, identify their problems and remedy them by applying wisdom and time tested techniques that have been experienced and explained by generations of global leaders and role models.

Today, in the knowledge-based world, personal mastery and excellence is essential to success, even survival. This requires practice of physical, mental, emotional and spiritual awareness. Self Science is all about understanding, choosing & living with these awareness.

Self science is a unique concept for personal transformation and is emerging with a renewed understanding in the 21 st century. Understanding of the principles of self science is needed not just at individual level, they are essential even at organizational and social levels. Similar shift is already happening in the field of science.

As Fritj of Capra writes in 'The Web of Life: A New Scientific Understanding of Living Systems', "By calling the emerging new vision of reality "ecological" in the sense of deep ecology, we emphasize that life is at its very center. Today the paradigm shift in science, at its deepest level, implies a shift from physics to life sciences."

We're at a tipping point in history. Thanks to the product of our rational mind and resultant technological skills, the last millennium has left us feeling dizzy and powerless with mind-boggling changes. Reading, Writing & Arithmetic have always been the foundation of our rational skills. But this millennium requires a different set of soft skills namely, **Spirit of Inquiry and Spirit of Awareness.**

These new paradigm form the foundation of **Self-Science Model which is explored and brought to the readers through this book**.

The intellectual saint, Ramana Maharishi says, "**Self Science** is an easy thing, the easiest thing there is."

The contents and expressions in the book can be explained with the following **Self Science Model**:

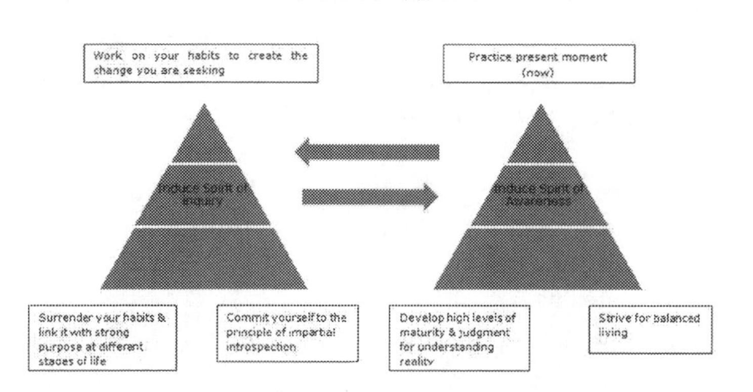

The book talks about two important pillars namely

- inducing the Spirit of Inquiry
- inducing the Spirit of Awareness

When the spirit of Inquiry is awakened, the individual has clarity of purpose at different stages of life. He is able to focus on his positive habits and erase negative habits for fostering success. He practices impartial introspection and is able to decide on the right choices.

When the spirit of awareness is awakened, he is anchored to present moment. As the individual develops high levels of maturity and judgment and understands reality he is in a better position to understand the moments of truth.

When the individual's spirit of Inquiry and spirit of awareness is awakened, it is easier to anchor to our true nature for balanced living. Human intelligence is insufficient to comprehend reality. Hence the need to anchor to one's true nature, which is not the body, breath, mind or intellect. In the Indian scriptures it is termed as**, "Satchitananda . . . .** Our true nature is knowledge, consciousness and bliss**."**

Now the challenge before all of us is to apply this Self Science Model at individual and organizational level.

In the beginning one may seem to be losing, but most successful people know that losing is only part of the game. If they lose, they realize that they need to work harder and become receptive to truth that comes from within. Losing is an opportunity to learn something additional-what didn't work and how it didn't work. When you have a sporting spirit and a play-the-game attitude and you lose, your response will be 'it is okay to lose'. Don't get upset. Just keep on trying again and again. Patience and Persistence will take over instead of pessimism and finally you will make it happen.

When you fall remember the goal is, "not to ever fall" but, "to rise each time you fall."

Many people who walked a path less traveled or heard the beat of a different drum are now getting connected as a force. It means that if you hunger for a deep change in your life that moves you in the direction of less stress, better health, lower consumption, greater spirituality, and more respect for Mother Earth and the diversity within and among the species that inhabit her, then you are welcome to practice **the self science model** mentioned in the book, and walk this toughest journey of life with a more integral world-view that is the synthesis of world and self.

Activity sheet

A) Plot graphically the peaks and valleys of your life events till now?

Few guidelines

1) Peaks are events in your life when you were high, happy, and prosperous and you have received more than you expected.
2) Valleys are events in your life when you were low-down, struggling and received less than you expected.

Pause, Reflect, and Scribble

Summary-

The Self Science Model

> *Common sense is the knack of seeing things as they are and doing things as they ought to be done.*
>
> *-Stowe*

- All living beings except humans know that the principle business of life is to enjoy day-to-day living
- We are best judges of all our thoughts, words and actions.
- This cyclic process of **thinking-action-reflection** has to be applied diligently in our life for progress and growth.
- Your quest will drive you to your purpose and unfold your potential.
- The world tramples all over us and pushes us down the moment we ask why . . . . Or why not . . . .
- Live a full life achieving your Purpose, unfolding your potential and realizing your wildest dreams . . . . !
- Now the challenge before all of us is to apply this Self Science Model at individual and organizational level.

CHAPTER 2

Induce the spirit of Inquiry

Ideas lose themselves as quickly as quail, and one must wing them the minute they arise out of the grass-or they are gone.

-Thomas F. Kennedy

Master rationalist Gautama Buddha, the founder of Buddhism said to his disciples, "Do not believe in what you have heard, do not believe in doctrines because they have been handed down to you through generations, do not believe in anything because it is followed blindly by many: do not believe because some old sage makes a statement: do not believe in truths to which you have become attached by habits. do not believe merely in the authority of your teachers and elders. Do not believe even in Buddha's teaching. Rather question your own thoughts, words and actions.

Swami Vivekananda also frequently refers, "Have deliberation and analyze, and when the result agrees with reason and is conducive for the good of one and all, accept it and live up to it.

To sum it up: No matter where you read it, or who said it, even if I have said it, **"Question it!"** And keep questioning till it agrees with your own reasoning and your own common sense. Thus continuously reinvent yourself with the wisdom attained from your questioning to suit the present circumstances. This will enable you to become, through action and reflection, subjects rather than objects of destiny. **You become the master of your destiny through questioning your own thoughts, words and actions!**

Swami Vivekananda had perfected this spirit of critical inquiry in his mind. A couple of days before his master Sri Ramakrishna Paramahamsa's passing, Vivekananda was seated near his bed. Seeing his Master's emaciated body suffering from excruciating pain, Vivekananda thought: Well, now if you can declare that you are God, then only will I believe you are really God Himself. Immediately Sri Ramakrishna looked up towards Vivekananda and said: "He who was Lord Rama and He who was Lord Krishna is now Ramakrishna in this body."

To change our quality of living we need to change our habitual questions to "quality" questions

Children are master questioners. As children we were learning only because of our curiosity to know through our continuous questioning, creativity and exploratory abilities. But as we grew older we forgot the art of questioning. We lost our curiosity to learn new knowledge, a new skill, or a new language. We became fossilized not ready to read new books, or try out and accept new ideas, and new friends. We thus became prisoners of our mind, boxed in our own little world, preferring stability and certainty, and as a result we clung to the status quo due to our excessive respect for the system and authority. Thus our ability to change stiffened. We need to blow away our perceptions, rules, prejudices and dogmas by our power of questioning for the magic to happen.

When we start echoing other's thoughts, dogmas and perceptions, we lose our originality and individuality. Successful people ask better questions and as a result they get better answers. It is quality questions that make all the difference. Sir Isaac Newton kept asking the question, why did the apple fall for twenty years to get the answer?

He went on to discover, Newton's Law of Gravitation which revolutionalized the history of humanity. Curiosity questions need to be defined well—then and there while seeking solutions to problems. Invite curiosity questions for getting insights. Seeing the invisible through curiosity questions is the key!

Leaders in the corporate, political, educational and social sector should encourage an environment which fosters spirit of Inquiry.

Pause, Reflect, and Scribble

____
____
____
____
____
____
____
____
____
____

Why is Spirit of Inquiry important for balanced living?

Spirit of Inquiry is self development.

To improve oneself & quicken our progress one must practice the spirit of Self Inquiry at least once a week. Preferably one should practice it as soon as one gets up in the morning or at night before going to sleep.

Ask yourself quality questions:

- How did I behave today?
- How was my attitude towards others?
- How productive was I?
- Did I accomplish the task and goals I set for myself today?

By reviewing these set of questions, one can identify the trends of our life and the directions in which we are moving. The questions that we ask can shape our perception of who we are! What we are capable of achieving! What we are willing to do to achieve our dreams! Often our resources are limited only by the questions we ask ourselves and others.

Warren Buffet, the great equity investor was once asked in a seminar, "I have $10,000; I want to double my investment by next year, where do I invest." Mr. Warren Buffet's simple answer was, Invest in **"yourself**."

Henry Ford also carried the same opinion, "Old men are always advising young men to save money. That is bad advice. Don't save every nickel. **Invest in "yourself"**. I never saved a dollar until I was forty years old."

Spirit of Inquiry can awaken the hidden potential in us.

This is the true story of the boy Venkatraman who was transformed to Bhagavan Ramana Maharishi, the world renowned spiritual leader from India.

In the year of 1892, 18th February Venkatraman was a boy of 12 years, when he sees his father's death. He sees the lifeless body of his father. The child wonders then and there about the nature of death. He inquires and practices Self Inquiry, Who am I . . . . ?

On 17th July 1896, four and half years later, Venkatraman is seated alone, in his uncle's home. He is suddenly gripped by an intense fear of death. He turns around and catches the fear of death face to face. The intensity and passion of the Inquiry turns his mind inward for the revelation . . . . !

In his own words expressed later, ". . . . When the body was dead and rigid, I was not dead. I was, on the other hand, conscious of being active, in existence. So the question arose in me, what is this **I** . . . ? Is it this body who calls himself **I** . . . . ? So I held my mouth shut, determined not to allow it to pronounce **"I"** or any other syllable. Still I felt within myself, the "**I"** . . . . what was that? I felt that was a force or current, a centre of energy playing in the body, working on despite the rigidity or activity of the body, though existing in connection with it . . . . the fear of death dropped off. **I,** being a subtle current had no death to fear".

The Inquiry was so intense, and so full of Awareness, Venkatraman got anchored to his true nature, **Satchitananda.** There was a permanent change in him from that moment. Practice of **Spirit of Inquiry** and **Spirit of Awareness** had liberated his **"I"** ness and **"My"** ness

The process of Self Inquiry is all about investing in "yourself."

It is simple and can be broken down into 3 parts.

Part One: Identification of what needs to change

(a) Identify the positive qualities one needs to develop.

(b) Also identify undesirable habits or flaws of character one need to get rid off.

Part Two: **Writing down and planning for the change**

(a) Write down the points and fix up a time frame for overcoming the deficiency.

(b) Every week pinpoint one quality or activity to be monitored

Part Three: **Monitoring and evaluation of action plan**

(a) From the list, choose one particular quality and work on it every day

(b) Watch over yourself and induce "spirit of awareness". Be ruthless with your observation.

(c) If you are flooded with negative emotion and are not able to control it, move away from the place of incident at the earliest and then try to reason. Our mind fails to accept our mistakes. For e.g. if our dad has told us to fill the petrol tank of the vehicle after using the vehicle, and we forgot about it. Assume he uses the vehicle the next day and is stuck midway without petrol. He comes home and shouts at us. On being questioned we start justifying why we did not fill petrol . . . . ? We give umpteen reasons, and the transaction ends in bitterness and conflict . . . . **Our powerful mind can reason our wrongs as right and we justify!**

(d) Begin and end the action plan with a date.

Thus by taking one quality at a time and integrating it with our day to day life, we can strive to make ourselves what we ought to be. Each individual step is significant and contributory.

A story is told about Chanakya the great teacher who build the Mauryan Empire in the year 325 B.C in India. Touring the countryside in disguise, he halted in a small village. An old woman offered him a meal. As Chanakya was hungry he accepted the invitation. He was served steaming hot rice. Chanakya delved into the centre of the rice on his plate, which resulted in his burning his fingers. "Oh dear" exclaimed the old woman, "you are indeed like our stupid minister Chanakya" Chanakya was taken aback. The old lady explained: "Never begin at the centre of the problem. Chanakya plans his attack on capital cities and loses the war. He should begin at the periphery and slowly make his way towards the centre." Chanakya had learnt a new technique for success. The beginning is always small and at the periphery. But it will unfailingly lead to the centre one day. **Start with small changes, path breaking changes shall follow later . . . . !**

Pause, Reflect, and Scribble:

____
____
____
____
____
____
____
____
____
____

Activity sheet

A) Worksheet for quality questions for personality development

Few guidelines

____

1) Am I in tune with nature?
2) Am I approaching my day with a beginner's mind set? And childlike approach?
3) Am I striving for self-control and purity in words, thought & action?
4) Am I free from stress, and sufficiently relaxed?
5) Am I providing service to society?
6) Am I really striving to be a happier and a better person every day?

7) Am I building my interpersonal relation with my near and dear ones?

8) Am I cheerful and enthusiastic?

9) Am I a role model in society?

B) Worksheet for writing the personal learning journal.

Note:- It may be advantageous to maintain a journal where we put down our thoughts & activities. It helps us in understanding, the direction we are moving. Also we can know more about ourselves by the behavior & attitudes of other individuals towards us.

Few guidelines

A personal journal can be a powerful learning tool for personal transformation as few can act according to their own inner will. I was exposed to this tool when I underwent training in T labs with the Indian Society of Applied Behavior. But a lot of us may be confused about what it is . . . . !

Effective journal keeping, as with other more private techniques (such as meditation), can be a means:

- For deepening our insights on the flow of our life,
- For learning to perceive and attend to the self in new ways,
- For discovering new dimensions of the self.

The human nervous system is so complex that a person can make **"himself"** happy or miserable, regardless of what is happening outside of himself, simply by changing his attitude and mind set. What we think, feel, see, and desire is information that we can manipulate and use. Thus we can shape the content of our lives, by our awareness and by developing a personal feedback system through our journal writing.

Reflecting on our personal experience can then be used as a feedback tool. **Experience is not what happens to man; it is what a man does with what happens to him**. We have an infinite amount to learn both from nature and our experiences. Each of us experiences life in a unique and worthwhile way. Experience is the hardest teacher; it gives you the test first and the lesson afterward.

Learning through personal journal is a lifetime process and one must develop this art from early childhood. I agree with what Francis Bacon says, "A man could do well to carry a pencil in his pocket, and write down the thoughts of the moment. Those that come unsought are commonly the most valuable, and should be secured, because they seldom return."

The key is using this art of writing down the thoughts that come unsought effectively for one's growth. However, some research indicates that our "staying power" is not as great as our "starting power". Being humans, we often lapse back to our old behaviors and unconscious habit patterns. Indeed, the phenomenon is common, reinforcing the idea that to work with and on one's own personal development may sound easy, but is unusually difficult to build into a consistent daily routine. Remind yourself that diamonds are only lumps of coal that stuck to their jobs.

'The Learning Journal'—an essay written by Margaret James Neill can be helpful in understanding journal writing. She mentions some of the possible content of one's journal and its potential impact on self. The potential effects can be achieved if one has a plan for using the journal. This includes writing, reading, and reflecting on your writing.

Your journal is your first step towards freedom and responsibility. The price of freedom is constant vigilance over yourself and your governance. Your journal can then become the vehicle through which you discover your own insight of the world.

It can help to expand your consciousness of the body-mind-spirit unity. It increases your sensitivity to yourself and towards others. **As we dig our own personal well, we may eventually reach the underground from which our basic unity springs.**

The motivation or energy we draw from Events, People, Feelings, Insights, Ideas and Dreams & Fantasies are different with different individuals. We need to create our own dashboards.

Check and Inquire on your trigger points:

One must analyze our life and learn from our experiences.

- View time and space as they come in the form of problems, experiences and relations.
- Look at the perpetual current of emotions and thoughts that arise, and develop high levels of intelligence, skill, wisdom, vision and love for work.
- Learn to be happier today than yesterday-that alone is the standard for measuring success.

Pause, reflect and scribble:

____

____

____

____

____

____

____

____

____

C) Worksheet for Quality Questions for Corporate Understanding

Few guidelines

---

1) What is the Business Model of the organization?
2) Are your employees familiar with the corporate functioning, of the different departments like marketing, finance, operations, human resource etc?
3) Are your employees using effective time management and team management principles?
4) Is the Corporate practicing business excellence?
5) Is the Corporate working on Business continuity Management?
6) Is the management providing "**WOW**" to its customers?
7) Do all the employees know the theory of Business?
8) Is the management developing competency of the employees?

Summary

Induce the spirit of Inquiry–

The life of every man is a diary in which he means to write one story and writes another, and his humblest hour is when he compares the volume as it is with what he vowed to make of it.

—James. M.Barrie

- To change our quality of our Living we need to change our habitual questions to Quality Questions.
- It is your Question which determines your thought'
- Children are Master Questioners
- Successful people ask better questions and as a result they get better answers
- Invite Curiosity Questions for getting insights and solutions.
- The questions that we ask can shape our perception of who we are! What we are capable of achieving! What we are willing to do to achieve our dreams! Often our resources are limited only by the questions we ask ourselves and others.
- That Nature is fair with everyone is a fact we do not realize. We need to blow away our

perceptions, rules, and dogmas by the power of questioning and introspection for the magic to happen.

- Self Inquiry can actually awaken the hidden potential in us.
- Analyze one's life as though in a mental laboratory. Examine the perceptual current of emotions and thoughts that arise within us, penetrate to the heart of aspirations, dreams, hopes and despairs. Dive deep into the mute cravings of the inner self. It is necessary to correct wrong thinking and bad attitudes. Hasten to develop positive habits for balanced living.
- To consider dispassionately one's fault is very difficult. But one can assess these shortcomings without developing an inferiority complex by using the personal learning journal profitably.
- Progress in life is very subtle. Never say you are not progressing. Develop self-confidence by conquering weakness. The greatest evidence of growth lies in deep inner effort to go upstream against one's old habits, and towards the source of lasting happiness. As we dig our own personal well, we may eventually reach the underground from which our basic unity springs
- Truly scrutinize your life. Find out what it really amounts to and then take steps to make it all it ought to be.

- Remember our staying power is not as great as our starting power.
- Every day try to be happier than the day before. That is the standard for measuring success.

References for further Reading:

Effective life Management by Swami Amartyananda

Self knowledge through Self Inquiry.—A golden jubilee publication of Bhagavan Ramana.

The Learning Journal—Margaret James Neill.

CHAPTER 3

Work on your habits to create the change you are seeking

> *Make it a point to do something every day that you don't want to do. This is the golden rule for acquiring the habit of doing your duty without pain.*
>
> *-Mark Twain*

When I started writing this book, the question very close to my heart was what made an individual successful? Is it right education, or right contacts? Is it being born in the right family? Is it right leadership? Or is it fate?

The search for an answer ended when I read the book written by Steven Covey, "The Seven Habits of Highly Effective People." It is a highly impactful book, the equivalent of an entire library on success literature.

This millennium requires knowledge and soft skills of a different type for success. The knowledge are the new 3 R's namely Respect, Responsibility & Rhythm. And Steven Covey has brought this out in his books.

The soft skills are **spirit of Inquiry and spirit of awareness** which is brought out in this book, Life is Balanced Living is Not!

Respect is re-spect, re-look at how we've understood who we are, where we're heading and how we are living. Respect leads to knowledge.

Responsibility is as Stephen R. Covey redefines, 'ability to choose response'. Responsibility leads to choosing rightly in every transaction.

Rhythm is all about joy, music, harmony and balanced living.

Covey also mentions in the book, "The Seven Habits of Highly Effective People." that one of his favorite essays is, "Common Denominator of Success" written by Albert E.N.Gray. I made a sincere attempt to locate and read this essay.

I would like to share the secret which Albert Gray discovered in his journey, **I Quote,** "Of course, like most of us, I had been brought up on the popular belief that the secret of success is hard work, but I had seen so many men work hard without succeeding and so many men succeed without working hard that I had become convinced that hard work did not lead to real success even though in most cases it might be one of the critical components . . . .

The common denominator of success-the secret of success of every man who has ever been

successful lies in the fact that he has formed a habit of doing things that a failure doesn't like doing.

The things that failures don't like to do are the very things that you and I and other human beings, including successful men naturally don't like to do. In other words, we have got to realize right from the start that success is something which is achieved by the minority of men, and is therefore unnatural and not to be achieved by following the natural likes and dislikes nor by being guided by our natural preferences and prejudices." **Unquote.**

I started applying the common denominators of success and it transformed my outlook towards life. Take for example in my teaching profession, To be a successful professor, one needs to read, research, contemplate and prepare a lot of material every time one goes into the classroom. But a lot of senior professors enter the classroom without adequate preparation. Therein is the problem. Routine sucks your energy and enthusiasm. One need to approach every experience which unfolds in our life from the beginner's perspective, curious and unsure of the outcome but yet prepared. Then there would not be even a single dull moment.

Very seldom do we realize that to attain health, and wisdom in life our success depends upon the battle between our good and bad habits. **We ourselves are the creator of our good or bad moments.**

One of the famous singers in India, Lata Mangeshker was asked the secret of her singing talent. She said, right from early childhood she had made a habit of getting up early morning for her practice session. "It was very difficult, when my friends were sleeping in the wee hours of the morning; I was at my practice session". She further added, "When I did not practice for a day, I was able to make out; when I did not practice for more than 2 days, my teacher was able to make out; and when I did not practice for more than a week—my audience was able to make out."

Mother's Bitter Lesson:

There was once a little boy who lived in a little Indian village with his rich and overindulgent mother. His father had died when he was very young, leaving him wholly dependent on the widowed mother's care. She was greatly attached to him, and she tried to fulfill his every desire. Being so blinded by her love, she thought he was perfect and could do no wrong, even when he misbehaved.

After a while the little boy began to attend the local village school. He was so mischievous that in a short time the whole school knew of this pest. The mother turned a deaf ear to the complaints of the teacher and the neighbors, and could find no fault with her boy—so pure in her estimate.

The boy grew worse day by day. He began stealing articles from his friends. One day he stole a costly silver bracelet from a classmate. He hurriedly got away from school and raced to the welcoming arms of his mother. She began to shower him with kisses and embraces; he could scarcely get her attention, so great was her affection.

At last, when mother's effusion of affectionate words ceased, the boy said, "mother, I tried to tell you, but you would not let me, I wanted the silver bracelet that my friend was wearing and so, cleverly, unnoticed, I picked it from my friend's wrist during play. The mother just gave the remorseful son a great big hug and said: You wanted it very much, and so I cannot scold you for taking it." The son was extremely astonished that he received caresses instead of the expected and well deserved scolding. Encouraged, the boy started stealing other costly valuables from his classmates.

Matters grew so bad that the village teachers called a special meeting of all instructors, and after heated discussions, voted to expel the spoiled child from school. The mother's intentions were all right, but due to her erroneous methods, the boy, as he grew older became a professional thief. When he grew to manhood, he became a criminal and joined a crime syndicate, scientifically planning and working out many crimes. The neighbors began to find their cattle and chickens missing, and other neighboring villagers began to lose their silverware and other valuable articles from their homes. There was great commotion, and vigilantes were appointed to apprehend the thief. The son, now an expert thief, became bolder and bolder because he was able to elude detection.

As no one can forever hide his wrongdoing, and no one can fool all the people all the time, the criminal ran into a trap laid for him by the outraged villagers. He was flogged by the angry villagers and remanded to jail. There was a hasty hearing and all the rich mother's attempts to save him failed. Her wicked son was sentenced to ten years of rigorous imprisonment for his confessed crimes of about fifty thefts.

As the son was being led to jail, he made a last request. He wanted to whisper a secret in his mother's ear. The mother came to her son, and placed her right ear in front of the son's mouth. No sooner was this done, than the son sprang and bit the ear of his howling mother with his saw—like teeth. After many punches and kicks from the policemen, he let go of his mother but only after biting off a piece of her ear. As the mother cursed and wailed, the son with great satisfaction in his eyes, cried out:" **Remember this bitter lesson;** if you had scolded me and corrected my habit when I stole the first silver bracelet of my friend, today I would not have to go to jail, and serve ten years of hard labor."

We must work on our habits to create the change we are seeking.

Pause, Reflect, and Scribble:

____

____

____

____

____

____

____

____

____

Why habits are important for balanced Living:

Habits are automatic mental machines installed by man to exercise economy in the initial use of will power and effort required for specific actions. Habits make the performance of such actions easier. Were it not for habits, one would have to learn anew every day how to wash, dress, eat and walk. The power of habits rules the day—to—day actions of the average person. A good many people let their whole life go by, making repeated good resolutions to improve, without ever succeeding in establishing and following new patterns of thoughts, habits and action.

More often than not, we find ourselves doing not what we wish to do, but what we are accustomed to doing. Every time you fail to do the right thing, you fuel the habit of doing the wrong thing

A person is a combination of nature and nurture. Our nature is our innate talent and the genes that we are born with. We are not left with any choice in accepting or rejecting them. But nurture is what is absorbed by training, and skill developed by practicing. It is within our power to mould and achieve the best in life through habit.

Habit formation is an important part of our nurturing. Habits are actions which are so internalized by frequent repetition that we have an involuntary tendency to perform them. We are creatures of habits, finding pleasure in what we have been used to doing. Once a practice becomes a habit, we can be sure that we are unlikely to lose it. So what we lack by nature, we have to supplement by nurture. So in the long road to success, do cultivate some important positive habits because they are powerful factors in our lives. **Only a changed man can bring about change in the world.** In due course of time, we start developing influence over our family members, business colleagues, community and social circles because of our strong, confident personality.

It is not only the "to do list" which is important, it is also important to adhere to your "not to do list" for your progress and success.

Pause, Reflect, and Scribble:

____
____
____
____
____
____
____
____
____

Activity sheet:

A) Worksheet for understanding how habits impact us:

Few guidelines

Management Game-

Find out what is it you do not like doing?

- At workplace
- At school/college
- At home
- At personal level

Capture your thoughts!

At workplace:

____

____

____

____

____

At school/college:

____

____

____

____

____

At home:

____

____

____

____

____

At personal level:

____

____

____

____

____

Debriefing:

- Thus we can observe, over a lifetime we have accumulated a lot of negative habits. But I personally believe these negative habits can be corrected.
- Just like habits can be learned they can be unlearned. It involves a process and a tremendous commitment and practice of the opposite good habit every day. It takes time to form either a good habit or a bad one.
- Change can be attained, by watching the quality of books we read, and quality of people who have an influence over us, our family members, our business associates, and our close friends. Also watch the food that you take in, as food directly has an impact on your thoughts.
- Once you are convinced that a certain habit is bad, starve it by avoiding all actions, circumstances, and persons that could stimulate it.
- Thus one can cure oneself of these bad habits by cauterizing them with the opposite good habits.
- Form right habits at the earliest stages of life to influence your future.
- Three thinkers: Albert E.N.Gray, Steven Covey, and Paramahamsa Yogananda have impacted me to have a critical look at my habits and work on them for my betterment. I would like to share few of their thoughts.

- Paramahamsa Yogananda refers to it thus, **"Bad habits are the worst enemies you can have . . . .** The habits that were formed earliest in your life have kept you quite busy until now; unwelcome habits have perhaps crowded out many worthwhile activities. The social wheel moves on the wheels of certain habits. You are punished by those habits. They make you do things you do not want to do, and leave you to suffer the consequences. You must drop bad habits and leave them behind as you move forward in life. Every day should be a transition from old bad habits to better good habits. Every year develop resolutions to keep only those habits that are for your highest good. The best way to get rid of your undesirable tendencies is not to think about them, not to acknowledge them. Never concede that a habit has a hold on you . . . . You must develop "won't habits." And stay away from those stimuli and thoughts that stimulate bad habits".
- Steven Covey says, "Because they are consistent, often unconscious patterns, habits constantly and daily express our character and produce our effectiveness . . . . or ineffectiveness. They have tremendous gravity pull-more than most people realize or would admit. Our character basically is a composite of our habits."
- Albert E.N Gray has referred, "If you do not deliberately form good habits, then unconsciously you will form bad ones. You are the kind of man you are because you have formed the habit of

being that kind of man, and the only way you can change is through habit. Every single qualification for success is acquired through habit. Men form habits and habits form futures."

Summary

Work on your habits to create the change you are seeking.

> *You're most brilliant ideas come in a flash, but the flash comes only after a lot of hard work. Nobody gets a big idea when he is not relaxed and nobody gets a big idea when he is relaxed all the time.*
>
> *—Edward Blakeslee*

- The secret of success of every man who has ever been successful, lies in the fact that he formed the habit of doing things that failures don't like to do."
- Very seldom do we realize that to attain health, and wisdom in our life our success depends upon the battle between our good and bad habits.
- It is not only the "to do list "which is important, it is also important to adhere to your "not to do list" for progress.
- Only a changed man can bring about change in the world.
- More often than not, we find ourselves doing not what we wish to do, but what we are accustomed to doing.
- What you lack by nature, you have to supplement by nurture.

- Habits are automatic mental machines installed by man to exercise economy in the initial use of will power and effort is required to execute actions.
- Were it not for habits, one would have to learn a new habit every day—how to wash, dress, eat, and walk. Habit, used rightly, can be a tremendous help on the path to freedom, but wrongly used, it can enslave the will almost hopelessly.
- It is within our power to mould and achieve the best in life through habit
- Just like habits can be learned, they can be unlearned. Men form habits and habits form futures.

References for further Reading:

1. **Steven Covey—The Seven Habits of Highly Effective People,**
2. **Steven Covey—The Eighth Habit**
3. **Steven Covey—Personal Leadership.**
4. **Albert. E.N.Gray-Common Denominator of success**

CHAPTER 4

Surrender your habits & link it with a meaning and purpose at different stages of life

What men want is not talent; it is purpose; in other words not the power to achieve, but the will to labour. I believe that labour judiciously applied becomes genius.

—Edward George Bulwer Lytton

I attended a leadership program last year, where an opinion poll was conducted among the participants. The results showed that 90 percent of the participants admitted that man needs 'something—for the sake of which he can live. Moreover, 70 percent conceded that there was something or someone in their own lives for whose sake they were even ready to die.

We were all stunned with the results. I was curious, I wanted to know whether Gen Y also thinks the same way. I repeated the survey at my University campus with 240 students and asked what they considered very important to them now at this stage of life, 25 percent of the students checked," making a lot of money":75 percent said their goal was, **"finding a purpose and meaning to my life".**

What struck me then was as teachers; we should not be hesitant about challenging students with, **"a purpose"** and help them achieve their best.

We are in an era where we are less connected with humanity, and have lost touch with our purpose. We have lost sight of the things that matter the most. The question a lot of successful people are asking is; how can I have greater meaning in my life? How can I make a lasting contribution through my work? How can I simplify life, so that I can enjoy the journey of life before it is too late? We should not sigh at our death bed, that, "I wish I was the person I could have been but never was".

There is much wisdom in the words of Nietzsche; **He who has a "why" to live for can bear almost any "how."**

There is nothing in the world, I venture to say, that would so effectively help one to survive even the worst conditions than the knowledge that there is a meaning in one's life. In the Nazi concentration camps, one could have witnessed that those who knew there was a task waiting for them to fulfill, were most apt to survive. The same conclusion has since been reached by other authors of books written on concentration camps, and also by psychiatric investigations into Japanese, North Korean and North Vietnamese prisoner-of-war camps.

Victor E.Frankyl also expressed the same thoughts in his book,' Man's Search for Meaning'. "When I was taken to the concentration camp of Auschwitz, a manuscript of mine was confiscated. Certainly, my deep desire to write this manuscript anew helped me to survive the rigors of the camp I was in. For instance, when in a camp in Bavaria I fell ill with typhus fever, I jotted down on little scraps of paper many notes intended to enable me to rewrite the manuscript, should I live to see the day of liberation. I am sure that this reconstruction of my lost manuscript in the dark barracks of a Bavarian concentration camp assisted me in overcoming the danger of cardiovascular collapse."

Thus it can be seen that mental health is based on a certain degree of tension, the tension between what one has already achieved and what one still ought to accomplish, or the gap between what one is and what one should become.

Such a tension is inherent in human being and is indispensable to mental wellbeing. What man needs is not a tensionless state but rather the pursuit of a worthwhile goal freely chosen. What he needs is not the discharge of tension at any cost but the call of a potential meaning-a purpose at different stages of life waiting to be fulfilled.

Why surrendering our habits and linking to a meaning and purpose at different stages of life is important for balanced living.

The life story of Chakravarti Rajagopalachari popularly known as Rajaji is an eye opener to us to surrender our habits and linking to a strong purpose. The greatest beauty of having **a purpose** is, it gives the ability to make yourself do the things you have to do; when you have to do them—whether you like it or not.

Rajaji was born on 8th December 1878 in a poor family in a South Indian village called Thorapalli. Though they were poor his father seeing the spark in him sponsored his higher studies completing Bachelor of law in January 1900. At the age of 21 he started his practice from Salem. From his first year he had a roaring practice and clients were paying him huge amounts for a case.

C.R as he was popularly known was not only professionally successful but he had time for friends, for recreation for politics, for social reform efforts etc.

Not only was C.R a reformist he was also a nonconformist taking on issues of untouchabililities, inter caste marriage, learning Urdu script and fighting against rules of orthodoxy. He set aside his practice to nurse his wife day and night when she was in deathbed and during the last months of her illness.

He was acknowledged as the top lawyer and would have been one of the wealthiest man in South India had he pursued his career. He gave up his flourishing practice and for the next three decades along with Vallabhai Patel, Jawaharlal Nehru, Rajendra Prasad and Abul Kalam Azad were the core of Mahatma Gandhi's political team for the struggle of Indian Independence.

He proved his leadership when he was made the Premier of Madras Presidency in the year 1937.As a premier he was innovative and levied sales tax on Tobacco, Petrol, and Power etc for the first time in Asia.

He was chosen as the first Governor General of Independent India in 1948 due to his extraordinary intelligence, clear thinking and sharp understanding.

Even after taking the highest position as Governor General of free India, for few years he rendered his services as the Central minister and also as the chief Minister which were lower position till he decided to retire.

He decided to nurse and built Swatantra party, when he felt that in a democracy there should be a strong opposition. He was already in his late seventies when he founded the party.

When he passed away at the ripe age of 94 in 1972 he was acknowledged far and wide as a great lawyer, freedom fighter, politician,reformist,teacher and writer. Rajaji has lived with a meaning and purpose at different stages of his life.

A lot many people take up a lot of resolutions in life and it ends up only as a resolution—nothing more than that. Any resolution or decision you make is simply a promise to yourself which isn't worth a dime unless you have formed the habit of making it and keeping it.

And you won't form the habit of making it and keeping it unless right at the start you link with a strong definite purpose that can be accomplished by keeping it. In other words, any resolution or decision you make has to be made again tomorrow, and the next day, and the next day and so on. And it not only has to be made each day, but it has to be kept each day, for if you miss one day in the making or keeping of it, you have to go back and begin all over again.

But if you continue with your resolution and keep it each day, you will finally wake up some morning a different man or a woman in a different world and you will wonder what has happened to you and the world you used to live in. In short I would like to conclude;

- Successful people have a strong purpose and similarly one can become master of self and master of likes and dislikes by surrendering your habits and linking it with a meaning and a purpose at different stages of life.
- With a strong purpose, our future is not going to be dependent upon economic conditions or outside influences of circumstances over which we have no control. **Our future is going to depend on our purpose in life**.
- All of which seems to prove that the strength which holds us to our purpose is not our own strength but the strength of the purpose itself.

- Develop an awareness of a meaning and a purpose worth living for at different stages of life.

Pause, Reflect, and Scribble:

____
____
____
____
____
____
____
____
____

Awareness of Meaning to life: Being in the present moment (Now) and with the flow of events.

The greatest problem with our population is the feeling of the total and ultimate meaninglessness of our lives. They lack the awareness of a meaning worth living for. To put the question," meaning to life", in general terms would be comparable to the question posed to a cricketer, "Tell me, Mr. Batsman what is the best shot to play?" There simply is no such thing as the best shot or even a good move, apart from situations like the condition of the pitch, condition of the weather, wear and tear of the ball, the personality of the bowler and of course the form of the batsman. The same holds true for meaning to life also

The meaning of life differs from man to man, from day to day, and from hour to hour. What matters, therefore, is not the meaning to life in general but rather the specific meaning of a person's life at a given moment, **the right way to live moment to moment.** The Taoist Scholar Chuang Tzu observed that the right way to live was to flow spontaneously, without hoping for rewards, yet moving with total commitment. This flow is characterized by a feeling that one's skills are adequate to cope with challenges in a goal oriented system, which along the way, provides clues to how one is performing.

Here I would like to reflect on the words of Robert Half, "The secret of what life is all about was answered by the sages; Life's about one day at a time no matter what your age is."

One should not search for abstract meaning to life.

Everyone has his own vocation or mission in life to carry out—a specific assignment which demands fulfillment. Therein he cannot be replaced, nor can his life be replaced. Thus, everyone's task is as unique as his specific opportunity to implement it. People who are successful generally enjoy what they are doing. Self registered goals, when concentrated upon to the limit, produce a flow in which new skills can be developed, psychic energy increased and actions integrated into a unified purpose. Purposes are fantasy unless translated to actionable goals. This is what I mean when I speak of a meaning to life.

Great People always inspire humanity.

The life of a great person, whatever his area of specialization, always inspires humanity. All great work is because of intensity of purpose.

Noah Webster spent thirty six years for creating the dictionary. He devoted his life to the collection and definition of words.

James Watt toiled tirelessly for twenty years on improving the power and efficiency of steam engines.

Sir Isaac Newton worked on his theory, law of gravitation persistently for twenty years till he shared it with the scientist community. It calls for great maturity, seriousness, and binding power to achieve your dreams. Great minds have purpose; where others have only wishes—Great people work on their purpose. This corroborates the thoughts of John Mennear,teacher and writer," Look at the stonecutter hammering away at his rock, perhaps a hundred times without as much as a crack showing on it. Yet at the hundred and first blow it will split in two, and it is not that the last blow did it, but all that had gone before."

Carry your compass with the needle pointing to your purpose.

When the compass was discovered, the ship captains used it in their ships in their journeys crossing continents. Even though many did not know why the needle pointed to the north they carried it to chalk out the sea route and reached their destinations and made the return journey successfully.

If you think about it, we all need the compass with the needle pointing to our purpose. As we go through our lives, we get distracted and defocused by the temptations around. We spend time watching movies and television shows, sleeping the extra hour and chatting nonstop.

And even though our parents teachers and our near and dear ones keep reminding us, we carry on with our bad habits The problem with many of us is we go through life with out a goal or a purpose.

Don't want to watch too much TV? Get yourself a book to read, clean and mop your floor.

Don't want to eat another pack of those chips? Eat a fruit, go for a plate of salads.

Don't want to become a couch potato? Start exercising, go for a jog.

So the best way to stop a person from doing something he shouldn't be doing is make him carry a compass with the needle pointing to his goal and purpose. The goals and purpose need not be long term it could specific, to be accomplished end of the day . . . . end of the hour!

Most of our problems in life start when we don't have anything meaningful to do. Having no purpose means not having to work towards achieving them. Not having a hobby or a passion means spending long hours idling away. And that old saying is still true. An idle mind is indeed the devil's workshop.

So starting today, get yourself **a purpose** that drives you out of bed every day.

Play a sport, indulge in a passion, and spend time on a hobby. Don't just sit there doing nothing. Living is all about translating our dreams to reality. What we all need is that compass. And that needle pointing to our purpose—one that excites us and keeps us on track. Once you have **a Purpose** you suddenly find yourself focused.

Find yourself a compass to carry. Find a purpose . . . . will you?

Pause, Reflect, and Scribble:

____

____

____

____

____

____

____

____

____

And how do we find our purpose?

Read books written by influential people and leaders. Meet them and make them your role models. Talk to them, and spend time contemplating on their thoughts and developing receptivity. You will get an answer. Small or big events, in the lives of great people, serve as a goal post. Depending upon our temperament and the capacity to imbibe, we can learn lessons like modesty, simplicity, forbearance, patience, rationality, practicality, purpose, passion and so on, from their lives.

Swami Vivekananda says, "As I grow older, I look more and more for greatness in little things. I want to know what a great man eats and wears, and how he speaks to his servants . . . . When we read their lives we do find extraordinary greatness even in small day to day events."

All successful people have their own limitations; some imposed by their background and upbringing, others by illness, yet others by their economic factors. But none of them allow limitations to become mental blocks. These are mental traps we set ourselves and get caught in: I am too young, I am too old, I am poor, I am rich, I am a woman . . . , I am just one person against the world, ... and so on.

We can turn any of our limitations into a mental block. But the special achievers among us demonstrate that limitations are meant to be understood and overcome.

Activity Chart:

A) Worksheet for understanding meaning and purpose at different stages of life:

Few guidelines

Objective: A group activity, which uses the concept of life cycle to assist participants to understand meaning and purpose at different stages of life.

Time requirement: 3 minutes for briefing

10 minutes for debriefing

1 minute per participant for the role-play

Overall requirement is 45 minutes.

Audience: participants are introduced to each other in an educational setting where the need exists for ongoing learning.

Learning opportunity: To promote discussions regarding the benefits of lifelong learning. Why it is both fun and important? For example, constantly changing work environments forces us to upgrade our skill sets constantly; there are also added benefits of Recognition of Pride Learning (RPL), for example overcoming barriers to lifelong learning (e.g. Closed Attitude) and some technique for overcoming self imposed limitations (e.g. By trusting and learning from each other)

Resources:

Give clear oral instructions; but printed instructions may also help participants.

Place chairs in a circle.

The activity can be conducted in a room with sufficient space for participants to move around and perform freely

Briefing the participants:

Form a circle and agree on a person to start. The participants are going to simulate lifelong learning. The selected person should behave as if he/she is 10 years old. The next person (clockwise) will act 20; the next 30 and so on. (If the group has more than 10 people one may consider smaller age increments)

The participant may represent a significant learning that suits the allotted age, for e.g.

A 15-year may learn a music instrument

A 40-year may be going back to University for upgrading particular skill sets.

An 80-year may need to adapt to new environment and forgo routine.

When participant comes forward he needs to state the allotted age, then he needs to show the significant learning situation or need for learning by telling it, showing it, acting it, miming it, drawing or even rhyming it.

Instructions to facilitator / trainer

Inform participants to form a circle and encourage the participants to metaphorically put on glasses with special filtering lenses and to look at the activity from the perspective of lifelong learning.

Start the group activity.

Debriefing the participants:

Begin with the statement that you like to "open the door" for participants to continue noticing lifelong learning and continue to trust the learning contributions from each other long after the simulation is finished.

Suggest that those in this group may bring expectations that training should be neatly packaged and laid on the table on a silver platter for them. They may bring their memories of secondary educational classroom setting and expect a repetition of this. Many may believe that the responsibility, for training &learning rests in the hands of their employer.

They are being given evidence that life is continual learning. They will need to become self-motivated and become lifelong learners to keep up with pace of change.

Company restructuring is creating management structures that are more efficient, with fewer positions for those who have climbed the corporate ladder. The concept of long-term job security is no longer a reality, unless one is meaningfully contributing to company's productivity. Staff is expected to be increasingly multi-skilled. **Lifelong learning is the process of planning and achieving personal goals.** Living and leadership is full of paradoxes. I would like to remind the readers of what Zimmerman says, "The universe pays every person in his coin. If you smile, it smiles with you in return. If you frown, you will be frowned at. If you sing, you will be invited in joyous company. If you think, you will be entertained by thinkers. If you love the world and earnestly seek good therein, you will be surrounded by loving friends, and nature will pour into your lap the treasures of the earth."

Some of the observations and experiences of the author in conducting the activity is as follows-

After a few role plays the group were able to do a better job, because they knew what was expected of them.

- The pace of life was brought out clearly / changes (physical, physiological, mental, and emotional) in life came out strikingly in those few minutes the participants played the management game.
- Reality of life also came out. Life is no bed of roses; it is full of happiness, sorrow, problems, stress, joy, health, and sickness. It is full of

duality, confusion, and conflicts. One need to passionately engage in daily routines and activities, without getting bogged down by them. Like the lotus, one should learn to live in muddy water but not get muddied.

- Getting into the role was difficult for some; for others it was exciting.
- Role clarity and the importance of various roles individual will play throughout lifetime were clearly brought out.
- Masking/unmasking of behavior patterns also comes out. Some roles were stereotypes. It was like playing the same old song.
- Having closed attitude and mental blocks were the reasons for not able to see reality. The choice is definitely ours. How we chisel our life depends entirely on us.
- Self-interest (personal gratification; happiness) versus group interest (maturity, commitment, loyalty) also was observed. Convey to the participants that secret of living is balancing the two namely self Interest and Group Interest.
- External environment can make us comfortable / uncomfortable, one need to practice equanimity to develop wisdom in life for only then does one realize how fleeting and precious life's gifts are, which once spent, can never be regained from death's unyielding grasp.

Wisdom from the research study by psychologist Havigurst will enable us to understand the life cycle learning much better. Havigurst has mentioned there are 6 critical life cycle changes which are important learning and development phases in our lifetime. Knowledge of life cycle prepares us for the next life cycle task and brings about awareness as to what the society expects from us. It also motivates individuals to do what the society expects him or her during the next life cycle. The insight of the research serves as a goal post and is as follows:

Babyhood and early childhood (till age 5 years)

- Learning to walk
- Learning to take solid foods
- Learning to talk
- Learning to control the elimination of body wastes
- Learning to relate oneself emotionally to parents, siblings and other people
- Forming simple concepts of social and physical reality

Late Childhood (5 to 12 years)

- Learning to work well with the peer group
- Becoming an independent person
- Learning an appropriate sex role

- Developing fundamental skills in reading, writing and calculating
- Developing conscience, morality and scale of values

Adolescence (13 to 19 years)

- Accepting one's physique and accepting a masculine or feminine role and establishing new relations with age mates of both sexes
- Gaining emotional independence from parents and other adults
- Achieving assurance of economic independence
- Selecting and preparing for an occupation
- Desiring and achieving socially responsible behavior

Adulthood (early—20 to 40 years)

- Getting started in an occupation
- Selecting life partner
- Learning to live with the life partner
- Starting a family and rearing children
- Managing a home

Middle Age (40 to 60 years)

- Achieving adult civic and social responsibility
- Assisting teen-age children to become responsible and happy adults
- Developing adult leisure time activities

- Relating oneself to one's spouse as a person
- Accepting and adjusting to the physiological changes of middle age
- Adjusting to ageing parents.

Old Age (above 60 years)

- Adjusting to decreasing physical strength and health
- Adjusting to retirement and reduced income
- Adjusting to death of spouse
- Meeting social and civic obligations
- Establishing satisfactory physical living arrangements
- Establishing an explicit affiliation with members of one's age group.

Note: Depending on the size and age group of the participants the facilitator can plan and bring out the intended learning in the de-briefing session. I will end this chapter with a powerful quote by John Cowper Powys, "If by the time we are sixty, we haven't learned what a knot of paradox and contradiction life is and how exquisitely the good and bad are mingled in every action we take; and what a compromising hostess our Lady of Truth is, we haven't grown old to much purpose."

Summary

Surrender your habits & link it with a meaning and purpose at different stages of life.

> *Most of life is routine, dull and grubby, but routine is the momentum that keeps man going. If you wait for inspiration you'll be standing on the corner after the parade is a mile down the street.*
>
> *—Ben Nicholas*

- He who has a "why" to live for can bear almost any "how"
- Teachers should not be hesitant about challenging, students with a purpose for them to fulfill, as everyone's task is as unique as his specific opportunity to implement it.
- One can become master of self and master of one's likes and dislikes by surrendering to one's purpose in life.
- With a strong purpose your future is not going to depend on economic conditions or outside influences of circumstances over which you have no control. Your future is going to depend on your purpose in life. All of which seems to prove that the strength which holds you to your purpose is not your strength but the strength of the purpose itself.
- Successful people have awareness of a meaning and purpose worth living for.

- One need to passionately engage in daily routines & activities of the day without getting bogged down by them.

References for further reading:

Victor Franklyn: Man's search for meaning in life.

100 inspiring lives:

Rajaji-A life: Rajmohan Gandhi

CHAPTER 5

Work on the principle of impartial introspection

> *One's life is not a straight line: it is a bundle of duties very often conflicting. And one is called upon continually to make one's choice between one duty and another.*
>
> *—Mahatma Gandhi*

The concept that we ought to know more about ourselves, goes back to the first time a human wondered, who am I? Whence do I come and whither do I go?

Socrates taught that to **"Know thyself"** is the basis of **"All Knowledge."** Shakespeare wrote, "To thine own self be true . . . . Thou canst not then be false to any man."

Like all the great ideas, the search for the concept of self realization rises and falls with the tides of history. But sooner or later every human embarks on this journey of self exploration. When thoughts of this nature awaken in the being, one needs to catch it and work on it patiently. When the quest becomes intensive and the student is ready, the teacher arrives. The teacher could be in the physical form of a personality, books, ideas or even an insight from your day—to day experience. Only after a patient struggle and the grace of the great spirits can you succeed in this endeavor.

Adi Shankara the Hindu saint who revived Hinduism in the year 2 B.C also emphasized the following, "Where have we come from? Who are our real parents? Where will we go when we die? We don't know. There are no answers. **Yet we need to ponder!"**

One is also reminded of the words Aurangzeb, the Great Mughal Emperor of India wrote to his son a few days before he died; "I came alone and I go as a stranger. I do not know who I am, or what I have been doing. I have been not the protector and guardian of the empire. Life, so valuable has been squandered in vain. I fear for my salvation. I fear my punishment. I believe in God's bounty and mercy, but I am afraid of what I have done. Every torment I have inflicted, every sin I have committed, every wrong I have done, I carry the consequences with me."

How does the self emerge?

The understanding of the self is a continuous journey, beginning with the emergence of self-awareness in the second year of the life cycle and gradually evolving to include the Self's characteristics and capacities through childhood and adolescent.

Babyhood

When the baby is born, it does not distinguish itself from the rest of the world. It cannot distinguish itself from the crib, the room and its parents. There are no boundaries, no separations. There is no identity.

But with passage of time and experiences, the child begins to experience the Self namely, as an entity separate from the rest of the world. When it has hunger pangs and is in discomfort, it requires its mother. When it is playful, Mother may not be available. The child then has the experience of its wishes not being its mother's command. Smiling and vocalizing at a caretaker who smiles and vocalizes back help specify the relation between the Self and the social world. And watching its own hand movements gives it another feedback—one tool which is under its direct control than other people or objects. Slowly a sense of separateness begins to develop. The child begins to have a separate identity—the **Self begins to emerge.**

Early Childhood

Self awareness quickly becomes a part of the child's emotional and social life.

At first, the child's sense of "Self" is so bound up with particular possessions and actions that it spends much time asserting its rights to objects.

In one study, two year's ability to distinguish between "Self" and "Others" was assessed. When each child was observed interacting with peers in a laboratory, the stronger the children's self definition, the more possessive they were about objects, claiming them as 'mine'. This was despite the fact that the playroom contained duplicates of many toys. These findings suggest that rather than being a sign of selfishness, early struggles over objects are a sign of developing selfhood, an effort to clarify boundaries between "Self" and "Others."

During early childhood, children begin to construct a self concept, or a set of beliefs about their own characteristics. These concepts are based on concrete characteristics, such as names, physical appearance, possessions and typical behaviors. These findings tally with psychologist Piaget and other theorists that acting on the environment and finding out what one can do provide an especially important early basis of self definition.

Once children became self aware, and were able to distinguish "Self" from "Others" there is an emergence of variety of emotional, and social competencies and skills.

Late childhood and adolescents

Over time, children seem to organize statements about **internal states and behavior** into dispositional conceptions that they are aware of and can tell others. Between the age of 8 and 11 yrs, a major shift takes place in children's self descriptions. They begin to mention personality traits. At first children mention overall qualities like, I am smart, I am honest, I am friendly. When these general ideas about the self are firmly in place, the adolescent start to qualify them as; I have a fairly quick temper, I am not thoroughly honest.

This trend shows that the adolescent understands that psychological qualities often change from one situation to the next.

Finally adolescent's self description play great emphasis on social virtues such as being considerate and cooperative. This is because they want to be liked and viewed positively by others.

What are the factors responsible for these revisions in self concept?

The content of the developing self is largely derived from interaction with others and what we imagine important people in our lives think of us. During middle childhood, children look to more people for information about themselves as they enter a wider range of settings in school and community.

Thus parents, siblings, peer group, teachers and society influence the Self concept. It is proven that children are not formless lump of hot metal waiting to be shaped; research indicates that children are born with styles of their own in the form of basic temperament patterns. As you may know from your own experience, infants and children differ greatly in this respect. Most are cheerful much of the time; others seem to be fussy and difficult to handle. Parental style may influence the way the child behaves socially, but parental style is also partly a response to the child's style. Parental behavior and child behavior influence one another in an ongoing cycle. **Finally it is not what you do for your children but what you taught them to do for themselves that will make them successful human being.**

Many researchers are of the view that two distinct aspects of the self emerge and become more refined with age.

1) The doer self.-The "visible I" The self is separate from the surrounding world. The self can act on and gain sense of control over its environment. The self has a private inner life not accessible to others. The self maintains a continuous existence over time.

2) The observer self. The "Invisible Me", is a reflective observer that treats the self as an object of knowledge and evaluation. It is the ability to stand aside, observe oneself without any prejudice, and judge accurately, it does not reason, it feels not with biased emotion, but with clear, calm intuition.

This tug of war between the **"I"** and the **"me"** is a continuous struggle and are intimately interlinked and influence each other. It is a universal struggle the one that rages daily in man's life from the moment of conception to the surrender of the last breath.

Paramahamsa Yogananda in his interpretation of Bhagwad Gita has shared thoughts about this tug of war, "Man has to struggle innumerable battles; biological, hereditary, bacteriological, physiological, climatic, social, ethical, political, sociological, psychological, metaphysics-so many varieties of inner and outer conflicts. The ordinary man often with his haphazard training has been found wanting in his understanding of this battle. The majority allow their characteristics to change passively and desultorily in various, undirected ways, according to their patterns of passing moods engendered by specific environment or according to the helpful or sinister influences of prenatal and postnatal habits . . . .

. . . . He alone who is **not "I" centered has the ability to see clearly and by practicing impartial discrimination can reduce the bewildering defeats everyday."**

It is the misuse of choices that lands us into problem.

As human beings we have infinite choices, but we are not aware of our choices. Normally we make a lot of choices in life unconsciously. After birth, the struggles of the infant are between its instinct to seek comfort and survival and the opposing relative helplessness of its immature body instrument. This confusion and struggle for making choices continues till our death bed.

Paramahamsa Yogananda says the story of every human is as follows, "As a child his first conscious struggle begins when he has to choose between his desire to play aimlessly and his desire to learn study and pursue some course of systematic training. Gradually more serious struggles force upon him from within or outside by bad company and environment. The youth finds suddenly, confronted with a host of problems that often he has been ill prepared to meet—temptation of sex, greed, money—making by easy and questionable means. Also pressures from his friends and social influences leave him confused many a time. The youth usually discovers he does not possess the wisdom or the will to face such worldly situations and experiences. As an adult he finds himself in a miserable condition, being overrun by money—making desires, destructive habits, failures, ignorance, disease and unhappiness which wreck his body and mind with scars of damage that gradually become irreparable and end in death."

Very few human beings use their power of free choices in making themselves the person they want to be. It is the right choice that makes the difference in our lives. **The moment you betray your good intentions, you slip!**

At the time of choosing, we know very well what is right and what is wrong, we also regret later of the compromise we may have made. Why is it then, that at the actual moment of committing the wrong action we do not hesitate? In spite of knowing everything, we are guilty of regrettable acts of violence, of indecency, of immortality, of corruption, of falsehood. **Why is it so . . . . ?**

- What is this dark force that compels us to indulge in such deeds even though in our saner moments we do not want to commit them?
- The answer is so plain, straight, direct . . . . and clear.

It is the inborn tendencies born of past actions. We have indulged in evil acts, repeatedly, and these impressions now become part and parcel of our nature. This is true whether we are discussing the urge towards lust or towards anger or greed which is present in all human beings as inborn tendencies. The remedial solution for these slippages is to work on the principle of Impartial Introspection.

Hence I would like to agree with what Wilfred A. Peterson says, "He was a wise who said; as I grow older I pay less attention to what men say; I just watch what they do."

Why impartial introspection is important for balanced living?

To fight the dark forces, one has to practice one's daily routine of introspection and overcome these urges. It is the greatest power in the world-to be anchored to our way of being through daily practice of impartial introspection. It is because we lack this practice that we often find ourselves miserable. It's a life time endeavor and very few ever make it a regular practice. **It is being rather than becoming, that one needs to learn.** The practice of impartial introspection enables us to wake up to life as it actually is, not the fantasy we often live in our mind.

I would like to narrate a popular story from the Indian Jataka Tales for understanding the thoughts of Dr Terry Warner's statement," Way of being is deeper and more important than behavior. Generally we respond to others way of being towards us rather than to their behavior. This means that others respond more to how we are regarding them, than they do to our particular words or action. Most problems at home, work and in the world are not failures of strategy but failures of way of being"

There were two friends named Mr.Ajay and Mr.Kannan living in an Indian village. They studied and grew up as friends and eventually became business partners. The small retail business which they started together was growing by leaps and bounds every year and in course of time was a popular mall in the village. Their honest trade practices and hard work was acknowledged far and wide.

Few years passed on; one fine day all of a sudden Mr.Kannan decided to wind off his partnership in the business and decided to go on a long pilgrimage. He went to his partner Mr.Ajay and conveyed his thoughts of squaring off his share in the business. Mr. Ajay tried a lot to dissuade Mr. Kannan from this decision. But Mr. Kannan stuck to his decision of relinquishing his share in the business and going on a long pilgrimage. Before starting on the journey he went to meet his friend Mr.Ajay and handed over five jars of tamarind in his safe custody.

Few days later Mr. Ajay's wife was trying to make, **"sambhar"** an Indian dish requiring tamarind concentrate. As there was no tamarind in the house, not knowing what to do, Mr. Ajay's wife asked him to buy some tamarind from the local shop. While he was dressing up, he suddenly realized that his friend Mr.Kannan had handed over few jars of tamarind which he had stored in the attic. He got a handful of tamarind from the jar and gave it to his wife.

When his wife was cleaning the tamarind during the cooking process, she suddenly realized there were some gold coins inside the ball of tamarind. She was overjoyed with her finding.

Immediately Mr.Ajay and his wife checked all the five tamarind jars and to their amazement they found hundreds of gold coins.

Mr.Ajay was thrilled. With this new found wealth he diversified his business operation. He created a chain of malls in the other nearby villages and cities. Many years passed and soon he became one of the wealthiest merchant in the region.

One late evening, there was a knock at his door. When Mr.Ajay went to check on the visitor, to his surprise, he found his dear friend Mr.Kannan.

I would like to stop the narration here and would like to have a dialogue with the readers.

Q) What are the responses/choices available to Mr.Ajay?

____
____
____
____
____
____
____
____
____

The reader's suggestions and solutions can fall in the following two columns!

Failing to recognize Mr.Kannan	Accepting Mr.Kannan and hugging him with tears of joy and warmth
Recognizing Mr.Kannan but confessing that the tamarind jars got stolen	Listening to his tale and parting with 50% of the wealth
Recognizing Mr.Kannan and handing over the tamarind jars without the gold coins stuffed inside.	Making Mr.Kannan his business partner
	Handing over the 5 jars of tamarind stuffed with the gold coins
"A" set of choices!	**"B "set of choices!**

Let us analyze the set of choices in this transaction around the following parameters: conflict awareness, thoughts, emotions & intentions arising in the character Mr Ajay a little deeply!

When Mr.Ajay makes," **"A set of choices,** what are the emotions going through him?

- ✓ Is Mr.Ajay worried about his actions?, Is he mean minded, Does he resents Mr.Kannan coming back?, Is he insecure, does he see Mr.Kannan as a rival, Is Mr. Ajay's nature controlling, manipulative, concerned with quantity, selfish, lonely, reactive, guarded, anxious, suspicious, fearful, rigid, defensive, self centered.

- ✓ Is Mr. Ajay's heart at war?
- ✓ Is Mr.Ajay treating Mr.Kannan like an object?
- ✓ Is Mr. Ajay's approach the resistance way?

When Mr.Ajay makes," **"B set of choices,** what are the emotions going through him?

- ✓ Is Mr.Ajay Interested and acknowledging Mr.Kannan? Is he abundance minded, delights in meeting Mr.Kannan? Is he secure, peaceful, sees Mr.Kannan as a friend, Is Mr. Ajay's nature trusting, sincere, concerned with quality, sharing, supportive, solicitous, open, assured and trusting, serene, flexible, accommodating and others centered.
- ✓ Is Mr. Ajay's heart at peace?
- ✓ Is Mr.Ajay treating Mr.Kannan humanely?
- ✓ Is Mr. Ajay's approach the responsive way?

I would like to share the following insights with the readers, **"Those who desire for peace must prepare for war . . . . inner war!"**

When our heart is at war our experiences are as follows:

- There is a rush of negative emotions, and one has to struggle, for being sensitive to the feelings of the other person.

- These negative emotions create confusion,
- We can't see situations clearly, we can't consider others' positions seriously enough to solve difficult problems,
- We have to fight for attentiveness and willingness for sharing of others' anguish.
- There is loss of memory and discriminative intelligence.
- We end up having poor interpersonal relationship with others.

But the irony is all the above mentioned experience happens so quickly in fractions of seconds the individual is not aware of this inner struggle and this inner war.

Can humans train themselves and overcome this inner struggles and this inner war . . . . ????

Pause, Reflect, and Scribble:

___

___

___

___

___

___

___

___

___

Activity chart:

A) Worksheet for making Choices?

Few guidelines

During any choice ask two important questions?

1. What are the consequences of this choice? (Check your intention, you will know in your heart. One's sense can never betray.)
2. Will the choice destroy my happiness and peace of mind or create pain to those around me? (If the answer is no, go ahead. If the answer is yes & it is bringing distress to those around us, don't make the choice)

Out of the infinite choices, only one choice will create happiness to you and everyone around you. Our ultimate goal is to move towards that right choice in speech, action and thoughts.

This knowledge is brought out in the Bhagavad Gita, one of the greatest epic written for humanity. The hidden meaning I feel in this epic is:

- To be aware of the continuous inner battles and inner struggles that is waged by humans throughout their lifetime.
- To Practice Impartial Introspection in every transaction for creating your own destiny.

Hence I would like to reiterate the clarity of the famous statement made by Swami Vivekananda, "By becoming conscious of your choice you create evolutionary actions. You create your own destiny, **Be and Make**."

B) Worksheet for daily practice of impartial introspection:

Few guidelines

Humans pack hundreds of activities everyday but one of the critical activities in life is, **"Impartial Introspection."** If you put fifty zeroes after one, you have a large sum. But erase the one, nothing remains. It is the one that makes many. Every night the power of one's introspection is to be invoked for reviewing the conflicts and transactions. Also determine on the outcome, whether favorable or unfavorable for creating evolutionary actions. As with the splitting of the atoms, the practice of Impartial Introspection gives us access to hidden power.

The young prince Siddhartha renounced the life of royal comfort and riches when: he saw a dead body ready for the funeral pyre; a person suffering from old age, and another crippled due to some disease.

Impartial Introspection changed him from Prince Siddhartha to Gautama Buddha. His teachings have continued to help millions of people get rid of their miseries all over the world for centuries.

The Impartial Introspection sheet lists all possible negative and their equivalent positive qualities that a human being can have.

Go through this list and at the end of the day, practice the game of Impartial Introspection for being at peace with oneself and with the universe.

Tick and work on the qualities for improvement every night before going to bed.

Ask yourself who won the battle between the two opposite camps.

What was the outcome . . . . ?

Impartial Introspection Sheet

Sr No	**Qualities**	1	2	3	4	5
1	Revengefulness/Forgiveness					
2	Pride/Humbleness					
3	Physical and mental abuse/Non hurtful behavior (self or to others)					
4	Self pity/resilience					
5	Showing off/non boasting, non bragging)					
6	Selfishness/unselfishness					
7	Swindling, deceiving(self or others) /Non deceiving					
8	Possessiveness/detachment					

9	Inquisitiveness/curiosity					
10	Spreading rumors (back-biting)/ non gossiping					
11	Being negative (thought, words or action)/ Being positive					
12	Materialistic/spiritual					
13	Cowardice (keeping quiet, no courage to speak up)/ Courageous and brave					
14	Lack of discipline /Discipline(thou ghts,words,action)					
15	Laziness/hardworking					
16	Nagging/understanding					
17	Being stubborn/ Being reasonable					
18	Unpunctual(not respecting others time)/ punctual					
19	Doing charity for publicity/Doing charity without personal gain and fame					
20	Jealousy/nonjealousy					
21	Deceitfullness/Truthfullness					
22	Hatred/lovingness					
23	Show of Anger/not losing temper					
24	Irritability/cheerfulness					
25	Hypocrisy/non hypocrisy					
26	Rudeness/kindness					

27	Depression or anxiety/non anxious					
28	Always complaining or grumbling/ non complaining or tolerant					
29	Greedy/self restraint					
30	Impatience/patience					
31	Keeping grudges/excusing offences					
32	Encouraging evil tendency/ Encouraging good tendency					
33	Avoiding responsibilities or duties/ taking responsibilities					
34	Sarcasm/sweetness in speech					
35	Impulse to choose wrongly/ Impulse to choose rightly					
36	Dominating/Accommodating					
37	Arguing unnecessarily/ nonargumentive					
38	Fighting/non bullying					
39	Worrying/non worrying					
40	Blaming someone else for your faults/Non fault finding					
41	Taking out your frustration on someone else/Being calm and peaceful					
42	Not practicing what you preach/ practicing what you preach					

43	Misguiding someone on purpose/ non misguiding					
44	Hurting someone's feeling/Being gentle					
45	Having fun at someone's expense/not poking fun at others expense					
46	Living beyond one's means/living within your budget					
47	Stingy/generous					
48	Wrong judgment/right judgment					
49	Not giving time for self Inquiry or self development activities/spending time on self Introspection					
50	Not controlling yourself mentally or physically/self control					
51	Manipulation / Non Manipulative attitude					
52	Restlessness/Being at ease with self and others(contentment)					

Note: Remember that the idea is not to be harsh. Life is like a grinding stone; whether it grinds you down or polishes you up depends on what you are made of. Be gentle and with the light of understanding dissolve the negativity and be receptive to the positive quality.

Practicing Impartial Introspection daily is difficult, but this practice will eventually lead to your healing guided by your inner light.

Pause Reflect and Scribble

Summary

Work on the principle of impartial introspection.

> *Get over the idea that only children should spend their time in study. Be a student so long as you still have something to learn, and this will mean all your life.*
>
> *-Harry L. Doherty*

- Parental behavior and child behavior influence one another on an ongoing cycle.
- He alone who is not "I" centered has the ability to see clearly and by practicing impartial discrimination can reduce the bewildering defeats everyday.
- It is the misuse of choice which lands us into problem.
- The goal of life is to practice impartial introspection to overcome our inborn tendencies and urges.
- As with splitting of atoms, the practice of Impartial Introspection gives access to our hidden power.
- Way of being is deeper and more important than behavior
- One's sense can never betray.
- Generally we respond to others way of being towards us rather than to their behavior.

- Other being respond more to how we are regarding them, then they do to our particular words or action

Those who desire for peace must prepare for war . . . . inner war!

- Most problems at home, work and in the world are not failures of strategy but failures of way of being. When our heart is at war
    - ✓ we are flooded with negative emotions,
    - ✓ these negative emotions create confusion,
    - ✓ we can't see situations clearly, we can't consider others' positions seriously enough to solve difficult problems,
    - ✓ there is loss of memory and discriminative intelligence
    - ✓ We end up having poor interpersonal relationship with others.

References for further Reading:

Dr Terry Warner-Self deception and leadership,

Dr Terry Warner-Anatomy of Peace,

CHAPTER 6

Align passionately to Present Moment (Now) by practicing awareness and silencing the mind

The best place to succeed is where you are, with what you have.

-Charles M. Schwab

In the deserts of Rajasthan, a caravan with several camels was moving from one place to another. The caravan stopped at a water hole for the night. Everybody was relaxing and an attendant to the head of the group said, "Sir, I have made a grave mistake." The Master asked, "What happened?" "I forgot to pick up some of the ropes to tie the camels with." The master reprimanded him for his negligence, and then said, "OK, we shall arrange for new ones tomorrow. But for today, just tie them without ropes."

"But, sir, how do I tie them without ropes?" he asked. The master replied, "You know there are no ropes but the camels do not! So, they think that you are tying them as you normally do. Go to them and pretend to tie them up." The attendant could not believe it but the master went with him and in front of every camel, he acted as though he was tying the rope. As the camel saw the action of being tied, it settled down on the ground.

The next morning the attendant came running to the master and said, "I think there is something wrong with the camels! They didn't get up when I went to them and tried to take them away. They must have some type of health problems."

The master asked him, "Did you untie them?" The attendant was confused, "But we didn't tie them." The master explained that the camels thought they were tied, so that notion needed to be removed. The master went to the camels and pretended to untie them and the camels jumped to their feet!

Like the tied camels, we live in an illusion of the past or the future. We are worried and stressed of the past actions and our poor relationships. We are comparing our wealth, qualifications, status and possessions with our known friends, relatives and siblings. We become so obsessed with our past mistakes or our lost opportunities that we are perpetually unhappy with the present situations. We are depressed and upset all the time trying endlessly seeking security and permanency. So in our mind we move on to the next flow of thoughts that is scheming, planning, strategizing about our future or continuously dreaming of a make belief world. Thus on analyzing, we find our mind keeps on planning and strategizing. It is busy either in the past or future or fantasy world so much so that in the bargain we miss the present. The irony is we usually do not see reality in front of us. We see life through our thoughts and concepts and we mistake our mental maps as reality. We are caught up so much in our mental world and engrossed in our activity and pursuit of pleasure and gratification that **present Moment (now)** flows by unnoticed, untouched, and untasted.

To bring out the concept of **Present Moment (Now)** I would like to share the following story.

Alexander the Great Greek Emperor was returning back home after conquering the world when he met the Great One in tattered clothes sleeping on the shores of Indus River. Out of pity, he got down from his horse and asked the sleeping hermit what he could do for him. The sleeping hermit opened his eyes and replied, "Step a little on the side, so that the sun's rays can fall on my body".

The watch word of the wise is **"Present Moment (Now)!"**

Ralph Waldo Emerson has beautifully said, "One of the illusions of the life is that the present hour is not the critical decisive hour. Write it on your heart that everyday is the best day of the year."

Developing common sense, leaving no unfinished work for tomorrow, and practicing equanimity in our day—to—day living without getting carried away by either our head or our heart in our transactions; is the key for balanced living.

Humans try to seek permanency in an impermanent world.

Every human being has a trait of constantly trying to arrange the world or people so that one is at ease and comfortable. We are stressed and dissatisfied with what is happening to us now. We want to change the present moment (now) if it is causing dissatisfaction and pain. For e.g. changing jobs, changing vehicle, or changing house or simple things like, turning on air conditioners or heaters to cool or warm up the room temperature, changing wall papers to suit our taste, or choosing our friends, co-workers and intimate partners.

- But people, things and situations do not remain the same.
- Our partner and co-workers lose their temper,
- The precious antique gets lost, stolen or broken.
- The new environment is not conducive.

The problem is we view impermanent things as permanent; though everything is changing around us. The process of change is constant and eternal. As you are reading this book your body is ageing, the book is decaying the print is fading, the walls of your room are developing cracks. You pay no attention to all of this. Then one day you look around you, your body is old and wrinkled, the book is yellow and a useless lump and the building has caved in. So you pine for lost youth and cry when your possessions are gone. When you ask yourself why the pain? The answer is plain and simple. It comes from your inattention. You failed to look at life closely. You failed to observe the constantly shifting flow of the world. You set up the mental construct of 'my body', 'my book' 'my property' and assumed that they were solid real entities. You assumed they would endure forever. They never do. So stop seeking permanency in an impermanent world. The key point to remember is much of the hurt moments in life disappear and many happy moments will emerge, if we learn to accept things as they are despite changing conditions. By developing awareness and attention you can perceive life as an ever flowing movement. You can take joy in the perpetual passing away of all phenomena. You can learn to live with the flow of reality rather than running perpetually against it. The insights gained from the practice of awareness can enable us to be in present moment (now). The key for transformation in life is to align passionately to present moment (Now) by practicing awareness and silencing the mind.

About inducing spirit of Awareness and silencing the mind.

Many of us spend our entire life with our body doing one thing while our mind is somewhere else e.g. while driving, while exercising, or while watching television. This type of habit brings in boredom, leakage of energy and high stress levels as our mind is never at rest. **Our mind is restful only when it is in the present moment (now) and is with the flow of events.** Awareness is all about paying attention to what we are doing with one pointed focus and to notice when we are not paying attention. Awareness unifies body, heart and mind bringing all of them together in focused attention for the magic to happen. True awareness is so intense, concentrated and finely tuned that one is able to pierce the inner workings of reality itself. Awareness is all about learning to look at each second as if it were the first and the only second in the universe.

Awareness is about observing! It is nonjudgemental, noncritical, nonthinking! In the process of observation what normally happens is the squandering of our attention:

- To cognizing the perception,
- Labeling it,
- And getting involved in a long string of symbolic thought about it.

That original moment of awareness is rapidly passed over due to this fragmentation of attention. One need not figure out everything. Concepts and reasoning just gets in the way. Don't think. Just observe and be with the flow, here and now. Be aware of the changes in thoughts, feelings and various emotions rising, do not try to control or hold on to it. Your relation with time and space should never be one of past or future but always of the simple and immediate now. Even after long practice if you find yourself suddenly waking up, realizing you have been off track for some time, don't get discouraged. Realize that you have been off track for some length of time. There is no need for negative reaction. The very act of realizing that you have been off the track is active awareness. Awareness does not aim at anything. It just sees! There is no struggle only acceptance.

Practice of Awareness opens up the understanding to experience life as it is—moment to moment. Awareness requires mental training and this practice develops the insight of reality and understanding of life. Practice of awareness enables us to see things as they really are, there is no dull moment. There is always a sense of excitement and wonder. These are wow experiences.

Maslow, who was a pioneer in studying the positive aspects of the human personality, gave a classic description of these experiences. Those experiences when we are completely aware are what psychologist Maslow calls **"Peak Experiences."**

"These moments were pure, positive happiness, when all doubts, all fears, all inhibitions, all tensions, and all weakness were left behind. Self-consciousness was lost. All separateness and distance from the world disappeared . . . ."

Although such experiences are rare-Maslow termed them, "peak experiences" for that reason—they have a curative power that goes far beyond their brief duration, which may be few days or few hours.

Maslow records that two of his patients, one a long term depressive patient who had often considered suicide, the other a person who suffered from severe anxiety attacks were both immediately and permanently cured after spontaneously falling into such experience (for each it happened only once)

Maslow was never able to give anyone else this peak experience-yet he was fascinated by these events that transcend normal life.

Awareness enables us to choose and respond moment to moment rather than react to situations. With practice of awareness we are in harmony with life in this universe.

Dr Deepak Chopra in the book, 'Quantum Healing' says, "The basic understanding that most have about themselves comes from thinking and feeling. This seems natural. The psyche of the individual is just not the head and heart, but also the field of silent intelligence. **The flow of thoughts and flood of emotions acts as a screen to keep the silence hidden**. We know very little about the field of silence and how it exercises control over us. Finding the silent gap that flashes in between our thoughts seems relatively easy but because it flashes by, a tiny gap is not a simple doorway, it forms a barrier no steel door could possibly match."

The gap is the field of silent intelligence which is not energy or strength, genius or insight, but it underlies all of these. It is life power in its purest form. The whole story of transformations of mind and body is contained in the difference between active and silent intelligence.

The practice of awareness could be providing a trip across the gap to delve deeper vertically in this field of silent intelligence. The mind experiences silence, wherein there is suspension of thoughts, emotions, drives, wishes, fears, ... everything. This practice is curative, intimate and supports life. Afterwards, when the mind returns to its usual pith (level of consciousness), it has acquired a little freedom to move.

The Silencing of the Mind model (2) could be explained as follows:

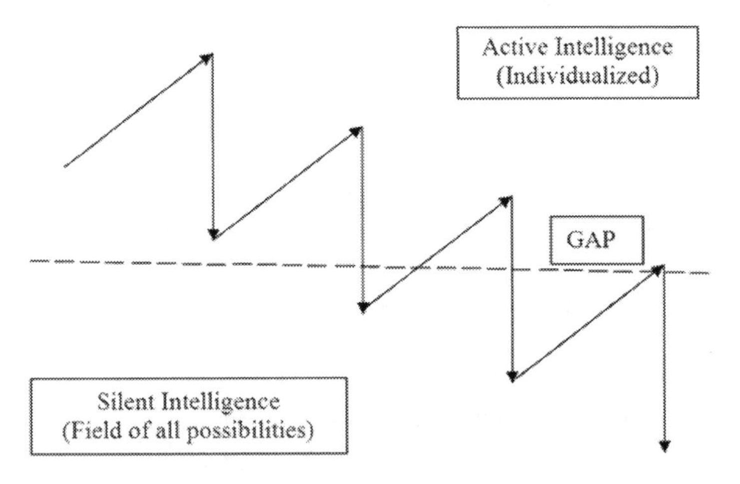

I feel delving into "peak experiences" transforms the animal mind to human mind thence intuitional mind and finally to spiritual mind. Brief insight of animal mind, human mind, intuitional mind, and spiritual mind is as follows.

Animal mind is driven by the forces of nature (internal and external).Whenever desires come they fulfill them; whatever emotions come they allow them to play, whatever physical wants they have, they try to satisfy them.

Human mind has reasoning and will but in many cases this control is very partial for reason is often deluded and flooded by vital desires and emotions and tries to justify wrong actions by wrong ideas, wrong reasonings and wrong movements. Here I agree with writer J.C Collins, "Half our mistakes in life arise from feeling where we ought to think and thinking where we ought to feel."

Intuitional mind is stirred by practice of awareness. The silenced mind learns to upgrade itself by "transmutation".It is only by persistence and continual efforts one is able to repel the gravitational pull of the lower nature.

Spiritual mind is the self actualization of the human mind in expressions of ideas and knowledge in life, not only in speech but in any form it can express.

To solve the mystery of the gap we need to consult and take feedbacks from the great ones who have dived deeper and found new realties. The paradox is that the skill to dive inside comes through practice of awareness and silencing the mind. It cannot be learnt otherwise.

Saint Beauve has nicely put forward, "I am only fulfilled, when pen in hand, I sit in the silence of my room."

Why building awareness and silencing the mind is essential for balanced Living:

The mind cannot be easily gripped; it is like mercury. Mercury cannot be held in the hand, it can be stored only in a capsule. The body is a capsule in which the mind is bound. As mercury spreads when spilled, our mind is spilled and diffused through our senses, thoughts and emotions. On deeper analysis we find, our mind is not steady it keeps on wavering and manifesting itself in different forms.

Thoughts, Emotions and Energy are manifestations of the mind.

The whole world is like a thermostat, where there is no loss of energy or water. The total quantity of energy or water remains constant in the universe, it only changes from one state to another state. Energy keeps moving from potential to kinetic and back to potential. Water keeps moving from ice to water to vapor and back to water.

Similarly it is believed that each individual being enters life with a store of energy. The energy is stored at the base of the navel region in humans.

There is a reference of this energy store in the ancient Indian Vedic literature. This subtle energy store releases small quanta continuously from birth to death. The quantity of release is fixed for a normal man depending on his spiritual growth. Some portion of this energy (Ep) is spent for various activities in the human system to maintain life. This energy spent (Es), will include the energy needed for basic autonomic activities (Es2), volitional physical activities (Es3), sense pleasures (Es4), and also for thinking and emotions (Es1).So energy expenditure equation (Es) is

$$Es= (Es1+Es2+Es3+Es4)$$

The remaining energy reaching the upper regions (E1) is the one that loads the head. Greater the energy E1 resonating in the head, heavier the head region.

The energy equation is,

$$Ep=Es+E1$$

The moment the energy force E1 starts hitting the conscious centers in the brain region, we immediately feel miserable and direct our energies in various activities-physical, mental and emotional. We eat away the energy, sing away the energy, walk away the energy, talk away the energy, and think away the energy, thus keeping the brain region with lesser loading. Greater the energy drawn away from brain centers, lesser the loading. It is our experience that if a normal man is asked to sit quiet without external aids, he would either sleep off or would feel miserable or restless. Modern man with greater sensitivity would burst himself out. Today the greatest punishment in judiciary is solitary confinement.

Thus it is observed that lesser the loading in the head region greater the lightness, well being, and happiness of the individual.

Our five sense organs have much greater energy draining capacities due to their subtle nature than the grosser physical instruments of action. At the point of contact of senses with their choice objects, a resonance is produced and energy flows profusely emptying the head region to the point that we are incapacitated to think.

So the solution for stress free and healthy living is to consciously flow the energy stuck in the head region productively within the frame of the body, and outside where boundary is no longer clear. When your energy melt and merge with the cosmos, we become boundaryless and totally alive.

The trick is to see that the energy becomes more fluid, unfrozen, melts and goes back to the cosmos. So if you are spending the whole day brooding and thinking it shows that you have more energy than you use for living; you have more energy than your so-called life needs.

In the day-to-day living our mind is always focused outside through our senses. There is a lot of chatter in the mind. It would be nice if we could just murmur that we must keep our mind peaceful and it magically happens, but for most of us, we must practice some activities for regulating the energy flow in the head and body and to make the mind peaceful.

Do not be miserly with your energy, fresh energy will be flowing. Let today be complete unto itself. Tomorrow will take care of itself. Don't be worried about tomorrow. The worry, the anxiety, the identification the problem all simply shows one thing—we are not living in the present moment. The key is to drop all fear, all identifications with our possessions, the work we do social status, and recognition, knowledge and education, physical appearance, special abilities, relationships, personal and family history, belief system, political, nationalistic, racial, religious and other collective identifications.

Here I would like to make a powerful statement, **"The concept of who you are restricts your growth or excellence."**

Pause, Reflect, and Scribble:

____
____
____
____
____
____
____
____
____

Activity sheet

A) Worksheet for understanding our true nature:

Few guidelines

- Pen down a transaction wherein, I betrayed the right choice?
- Or in the same transaction, I honored the right choice?

Debriefing:

The world may praise us, but we can never cheat our consciousness . . . . We know our true worth. We are best judges of all our thoughts, words and actions.

Human intelligence cannot comprehend reality. Hence the individual have to anchor to their true nature which is not the body, breadth, mind or intellect. In the Indian scriptures it is termed as, "Satchitananda . . . ." Our true nature is Knowledge, Conciousness and Bliss.

By anchoring to our true nature the individual can blossom to the fullest and be empowered in health, wealth and success in relationship at home and at the workplace.

Perhaps it is a good idea to befriend, cultivate and strengthen our true nature, **"Now"** at the earliest rather than to wait until our deathbed and lament.

Pause, Reflect, and Scribble:

____

____

____

____

____

____

____

____

____

Summary

Align passionately to present moment (now) by practicing awareness and silencing the mind.

> *A moment's insight is sometimes worth a life's experience.*
>
> *-Oliver Wendell Holmes*

- Mind is busy either in the past or the future or a fantasy world so much so that in the bargain we miss the present.
- Humans try to seek permanency in an impermanent world.
- Much of the hurt moments in life disappear and many happy moments will emerge, if we learn to accept things as they are despite changing conditions.
- Mind is not steady, it keeps wavering or manifesting as thoughts, emotions and energy.
- Our mind is restful only when it is in the present moment (Now) and is with the flow of events.
- Awareness unifies body, heart, mind, bringing them together in focused attention.
- The Energy equation is Ep=Es+E1,the moment the energy force E 1 starts hitting the consciousness centre in the brain region, we feel miserable.

- Practicing equanimity in our day –to—day life without getting carried by either our head or our heart in our transactions is the key for successful living in the present moment.
- The watch word of the wise is Present Moment (Now)!
- The flow of thoughts & flood of emotions act as a screen to keep the silence hidden.
- The concept of who you are restricts your growth or excellence.
- Perhaps it is a good idea to befriend, cultivate and strengthen our true nature**, "Now"** at the earliest rather than to wait until our deathbed and lament.

References for further reading:

Eckhart Tole-Present Moment,

Eckhart Tole—Silence speaks,

JD Krishnamurthy-Awakening Intelligence.

Swami Vivekananda volumes

CHAPTER 7

Develop high levels of maturity, and Judgment for understanding reality

Men occasionally stumble over the truth, but most of them pick themselves up and hurry off as if nothing happened.

-Winston Churchill

Humanity in general is soaked in mediocrity and hence loses sight of reality. It is so important to know what to take seriously, and what not to take seriously, especially when one interacts and communicates with others. A truth very hard to digest is a lot of individuals land into serious trouble in life, if they start believing and working on "not to be taken" things seriously and implement it.

Everything we perceive in existence is a combination of subject and object. If "one" changes so does the "other." And since each of us, as subject or object, is different and unique, we perceive the same world differently. Truth or reality is what exists but due to the baggage we carry we are not able to see the truth as it exists. Our baggage is our memory, feelings, thoughts, perception, interpretations and reactions which we add to this reality.

If reality is the salad, our baggage becomes the pepper, salt and vinegar. If the salad forms the major portion and the pepper and salt form a minor one, the serving would be palatable. But if our toppings are more than the nature of the salad, the salad would taste terrible, and would be unfit for consumption. This is what is happening to most of us; **we tend to add our baggage (perceptions or interpretations) to every event that occurs in our lives.** Perhaps that is why we have more unhappiness and sorrow, since we read too much into reality and get depressed with what is happening to us and around us most of the time. Thus a lot of resources, time, relationship and energy go down the drain if we do not develop high levels of maturity and judgment for understanding reality.

We see this world through our mind and ego.

The daily transaction we experience is filtered through our mind. Every word, dialogue, interaction that we encounter is analyzed by our mind. And since no two minds are alike, my interpretation of transaction is totally different from yours or anyone else's. The reason why people don't see eye to eye is simple, for it is rare for two persons to see things in exactly the same manner. The more silent and blissful the mind is, the less judgmental and egoistic we are, and greater the chance of us coming close to reality.

The degree of conflict that we have with others will be directly proportional to the amount of baggage we bring into any situation.

In order to achieve our goals, we try and change words, events, people, company and circumstances. We want to achieve materialistic goals without changing ourselves. Though sometimes we achieve them, many times we fail, because we often forget that everyone in this world is also doing the same. **The solution is to change from within where we have total control, than the outside world over which we have limited control.**

Mahatma Gandhi rightly says, "Be the change that you want to create in this world."

Any situation can change provided we are ready to change. By changing our bad behavior and dropping our baggage we shall be taking the path of acceptance and surrender and instantly we can transact from conflict to compromise, noncooperation to cooperation, and misery to happiness.

Swami Vivekananda has said that we waste two thirds of our energy. Leaders who have achieved the pinnacle of success in their careers—be they renowned doctors, great scientists, or accomplished musicians use, at most only one third of their energies. Just imagine all that we could achieve if we had access to even a fraction of the remaining two thirds! All our dreams and projects would then be within our reach. As the saying goes ' Genius is only the superior power of seeing and as average humans, we don't see things as they are; we see things as we are.'

We waste a lot of energy.

When we compare ourselves with others or complain about others we trigger negative emotions such as anger, fear, greed, shame etc. The immature mind compares with, what is and what should be . . . . ! Right from childhood, we are trained to compare ourselves with the topper of the class; our parents compare us with our siblings or with the neighbor's sons and daughters. As adults, we continuously compare, he is better than I, intelligent than I, smarter . . . richer . . . or she is more beautiful than I, brighter . . . . etc.

A person may get a coveted job with a good salary, but the moment he sees the neighbor getting more salary for the same job, he becomes depressed and gloomy and starts feeling miserable and angry. We often find persons complaining and grumbling at the drop of a hat, either about the weather, or about their organization, job, boss, colleagues, wife or their health. They are never content and satisfied with their present state of affairs.

This comparison, complaining, and categorizing way of thinking, is the beginning of wastage of energy. Our energy is also dissipated when we have negative emotions like anger, jealousy, anxiety, fear and despair. Thus it goes on and on; like an itch that cannot be cured with any amount of scratching. This comparative, categorizing and complaining way of thinking torments us till we reach our funeral pyre.

One's physical body is a regulator for flow of energy. If one is able to direct and channelize this flow of energy then one can create any amount of wealth and prosperity.

Most often our energy goes into comparing with others, complaining about others, categorizing our experiences and upholding our importance. If we were capable of losing some of the comparisons, complaints and categorization three extraordinary things would happen to us.

We would 'free' our energy from trying to maintain the illusory idea of our grandeur or hiding our fear, anxiety, anger and jealousy.

We will have enough energy to catch a glimpse of the reality of the present moment.

We would develop the skill of seeing, thinking, feeling, speaking and being with the flow, in every transaction.

So how do we experience reality for balanced living?

- ✓ Learn to accept things, circumstances and people as they occur without comparison, complaints or judgments. By being nonjudgmental about people and things, one is willing to drop one's ego and one's baggage.

- ✓ Take responsibility, including for **"yourself."** Responsibility means not blaming anyone or anything for your present circumstances.
- ✓ Stop defending your point of view. People spend most of their time defending their point of view rather than listening and engaging in a dialogue with others. Defensive attitude blocks the miracle from happening in the present moment.
- ✓ Live life with energy, drive, and participate passionately in every transaction moment to moment.
- ✓ Continuously cleanse and purify the mind

With these renewed understanding, living becomes balanced with an exquisite combination of acceptance, responsibility, defenselessness, and life brims with positive energy!

Pause, Reflect, and Scribble:

____
____
____
____
____
____
____
____
____

Activity sheet

A) Worksheet for cleansing and purifying the mind:

Few guidelines

As part of his routine, Lord Buddha used to go on his round of begging every morning. One day a milkman, deeply influenced by Buddha insisted that the Master visit him and share his wisdom, for the exchange of which he offered some milk. Buddha agreed. In the evening when Buddha set out he took with him a container in which he intentionally put some mud. The milkman was about to pour some milk into it, when he realized that the container had some impurity. So, he removed the impurities, cleaned the container, poured the milk and returned the container to Buddha. Upon receiving the container, Buddha got up to leave. The milkman ran after him and asked him why he was leaving without imparting any wisdom. Buddha replied he just had imparted it. Not able to fathom the learning the milkman requested Buddha to explain. The Master then explained that the mind is like the container and the thoughts that preoccupy it are like the impurities found in the container, "To attain wisdom one must purify the mind by making it free of all impure thoughts. Only by cleansing the mind shall we be able to imbibe any further learning."

We came into this world to free ourselves from the baggage we carry but we got entangled more and more with our bad habits, evil tendencies, our ego and our negativities. **We are spiritual beings born for experiencing humanness**. We need to cleanse and purify our mind to regain our glory and birth right. Cleansing is all about emptying our past impressions and contents of our mind. It is all about dropping our ego, and the baggage we carry. Zero state of mind can be attained only by constant alertness. **Our mind cleansed of all impurities, is a powerful tool of unlimited potential.**

Case study:

This is a true story shared by Joe Vitale in the book," Zero Limits", of a unique cleansing process used by Dr Hew Len, clinical psychologist when he was employed in the services of Hawaii State Hospital. Dr Hew Len healed an entire ward of mentally ill criminals not by using drugs, medication, counseling therapy, evaluations, assessment, or diagnosis. Nor did he use any psychological testing. He did not attend case conferences or participate in mandated record keeping. He instead practiced a **process of cleansing** i.e. working on himself, which had to do with taking 100 percent responsibility on yourself, and allowing the removal of negative thoughts and unwanted energies within you.

The philosophy being—whatever is happening to you or around you, you are responsible and the only way for impacting things positively is to cleanse yourself.

In the words of Dr Hew Len, "More than anything, I did my cleansing before, during, and after each visit to the ward, week in and week out for three years.

I cleaned with whatever was going on in me, with the ward, every morning and every evening.

And if anything about the ward came up in my mind, I started with the cleansing process. The cleansing process is all **about repeating "the four magical phrases;"**

- ✓ I love you,
- ✓ I thank you,
- ✓ Please forgive me,
- ✓ I am sorry,

This unique cleansing process, you and I can practice very easily. That's it. It may be the shortest route to success ever created. It might be the path of least resistance. It might be the most direct route for entering the zero state of mind. And it all begins and ends with the four **"Magical Phrases."**

Pause, Reflect, and Scribble:

Summary

Develop high levels of maturity, and judgment for understanding reality.

> *Spiritual and Social upliftment of humans and purity of mind are essentially the same thing.*
>
> *-Jain Teachings*

- In a comparative, measuring or complaining state of mind, we waste a lot of energy.
- The immature mind compares itself with what is and what should be.
- Our baggage is our memory, feelings, thoughts, perceptions, beliefs, interpretations, and reactions. Our baggage distorts the reality.
- The degree of conflict that we have with others will be directly proportional to the amount of baggage we bring into any situation.
- Cleansing and purifying our mind is all about dropping our ego, and the baggage we carry. Our mind cleansed of all impurities is a powerful tool of unlimited potential.
- Whatever is happening to you or around you, you are responsible for it and the only way for impacting things positively is to cleanse yourself.

- The most direct route for entering the zero state of mind begins and ends with the four magical phrases:
  - I love you
  - I thank you
  - Please forgive me
  - I am sorry.

References for further reading:

Joe Vitale-Zero limits

CHAPTER 8

Strive for Balanced living

Everyday . . . . the most important thing in human life is living it.

-Frank R Barry

In our present high tech era, we have all materialistic comforts and untold wealth and prosperity. But yet living is full of stress and tension. Is bringing heaven on earth a utopian concept?

We search for happiness by acquiring materialistic possession and satisfying our desires i.e.

- Ownership of external objects (money, dress, wealth, houses and property)
- Ego satisfaction (power, status, fame, etc.)
- Sensual satisfaction (sex, alcohol, drugs, good food, personal comforts, and sensual pleasures)

Even after having a lot of these materialistic possessions, happiness eludes him.

Why this state of delusion?

What is the answer in this quest for happiness?

The eastern philosophy provides us with an answer. It encourages us to realize happiness by showing us the process and also the methodology.

Insight of our true Nature awakens us from delusion:

Many scientists are now acknowledging existence of inherent intelligence in life and in creation. Physically each human entity is a group of over sixty trillion living cells that do what they do by themselves. Every second there are billions of activities happening inside our body that keeps us alive and we have absolutely no control over it. Blood is pumped, cells are created, and hormones are distributed throughout our body. A million things out of our control have to happen just for us to digest our food and we are not even aware that it is happening right now within our body. Similarly there is an exchange of information with other living entities happening simultaneously in this universe, silently, without our knowledge with clock like precision. In fact most people are not aware of the fact that they do not control what is going on outside of them either. Most people believe they are in control and the exact opposite is the truth, because they have no insight of their true nature. Pancha Kosha model from the Taitariya Upanishad provides us the insight of our true nature: "We are not the body, nor the life force, nor the mind, thoughts and emotions, nor the intellect either. Our true nature is bliss."

In our original state we are totally stress free and in peace. We are in a blissful state. Our true nature is pure knowledge pure consciousness and pure bliss-**"Satchitananda".** The fact is, one's true nature is completely free of "I ness" and "My ness". It is unfearful of any challenge. It feels inferior to no one. Yet it is humble and feels superior to no one. It recognizes that everyone else is the same self, made up of the same stuff, the same spirit or life spark in different disguises.

Ramakrishna Paramahamsa used to say, "Each life—spark is potentially divine. Respect life in all forms, whether in cats, dogs, birds or humans. It is our human perception which is at fault. We have graded mankind as superior to other species".

Humans need to understand the truth; Life and this universe is not a random event. Spark of life is perfect and need to be respected and honored in all forms! Relax and flow with the happening. The truth is so obvious and simple we need some one to express it simply, **"Life is balanced."**

We may call this state as a state of perfection. When this state of perfection gets disturbed in humans, there is imbalance, thinking starts, ignorance sets in, leading to further thinking and we start limiting ourselves. Ignorance leads to manifestations of **"I-ness",** ego and the associated attachments, **"My-ness"**—my mind, my thoughts, my feelings, my body, my possessions etc.

The problem is, as individual spark of life, we associate ourselves with our body, mind, ego, memory thoughts, status, positions and material possessions. The ego is our self-image. It is a social mask. It is the role one is playing. Our social mask thrives on approval. It wants to control and it is sustained by Power. It lives on fear, afraid of losing its identity, possessions and relationships. We identify with name, form, status, power, positions etc. All these disturbances and Imbalances leads to fragmentation of identity, crystallizing as personality. The individual personality start asserting strong likes and dislikes. It is then that emotional upsurges start. Tossed up and down in these emotional imbalances, large amount of energy is expelled. Speed increases in our thoughts leading to restlessness in our actions. Excess speed and restlessness brings further imbalances, impairs efficiency and leads to a deterioration in the quality of life we are living.

The journey of every spark of life is unique. **The purpose of spark of life is balanced living in human form.**

Balance doing with thinking. Balance head with heart. Balance firmness with being friendly and approachable. Balance conflict with peace. Balance compassion with courage. Balance making it to happen with allowing it to happen. Balance freedom by being spontaneous with also being responsible.

Hence the theme of the book, "**Life is balanced Living is not!**"

Pause, Reflect, and Scribble:

____

____

____

____

____

____

____

____

____

How do we strive for balanced living?

Practice silence.

Tune in to our true nature. This co-existence of opposites, stillness and dynamism at the same time makes us independent of situations, circumstances, people and things. This world is the combination of ever dynamic, ever in motion energy and the quiet, still eternal silent life—spark. Stillness is not void. It is full of creativity.

Practice being non-judgmental

Practice being non judgmental, as everyone in this world is unique, and has something to offer. Develop the attitude of equanimity in all situations. This attitude will enable you to silence your mind and there shall be no wastage of energy. Universe in its eternal state is energy and information, which is ever alive and eternal like thought. Universe is ever changing, ever dynamic reality.

Practice non-resistance

Non-resistance can be practiced by allowing the external world to impeach, penetrate through you, without becoming tense or stressed up. Our senses are externally focused. We seek the external world through our senses, and mind. The five senses help in governing the body.

Smell is the earth element.

Taste is the water element.

Touch is the air element.

Sound is the ethereal element.

Shape and form is the fire and light element.

The entire universe is made of these five elements (earth, water, air, ethereal, fire and light) which also exists in our body. That which exists in our body also exists in nature. We get pulled by the changes both in the internal as well as in the external environment. Hence we are in bondage.

The logic is if one has control over the internal nature one also develops control over the external nature.

Individuals who have control over these five elements (earth, water, air, ethereal, fire and light) has control over the body and hence the universe. The secret of living is being with the flow of life. Do not fret or worry! **Be in the present moment joyful, happy and contented.** The rose flower does not fret that others do not smell its fragrance. It just blooms and spreads its fragrance irrespective of others' opinion or reactions.

Practice what Horace says, "Dare to be wise; begin! He who postpones the hour of living rightly is like the rustic who waits for the river to run out before he crosses."

Practice yoga

Increase your three dimensional awareness by practicing yoga. One does not require any evidence or any one to tell us we are alive. Our awareness itself is our proof. Individual spark of life is the awareness within that cannot be destroyed by time or space. Cosmic consciousness is the unseen power which governs the universe. Getting anchored to cosmic consciousness is real yoga.

Activity sheet

A) Worksheet for practicing yoga, time tested technique for balanced living.

Few guidelines

In the 21 st century people find themselves frequently in a state of unhappiness and dissatisfaction in spite of rapid material progress. It seems people have become increasingly effective in managing their careers and social lives, but are equally ineffective in managing the most fundamental aspect of their lives, the ability to **attain peace of mind**. Many are successful people, brilliant in their respective fields but unsuccessful in building happy fulfilling lives.

The Western world found it extremely difficult and confusing to understand how mind and body interact with each other. This led to the practices of physicians concerned with the treatment of body, psychiatrists and psychologists working with the healing of the mind. The gap only widened due to lack of communication and understanding between the two camps.

Now the time has come to solve the conflicts between the two camps (Physicians and psychiatrist). Yoga practiced by ancient Indians, is a great piece of literature, throwing light on the body—mind complexity. The essence of Yoga literature is balanced living and provides solutions to all our physical and psychological problems. These messages are meant for the betterment of humanity. The teachings of Yoga has a universal appeal and are not the exclusive property of any particular nation or creed as they encompass every being, irrespective of caste, color, nationality, sex, race or religion.

These messages or understanding cannot be taught like history or mathematics; it is transmitted like light or heat.

The essence of Yoga has been defined very crisply by ancient Indian seers Patanjali, Vasistha and Sri Krishna as follows.

Yogah chitta vritti nirodhah .... **Patanjali**

"Yoga is mastery over mind"

Manah prasamana upayah.......................... **Vasistha**

"Yoga is a technique to calm down mind"

Samatvam yoga uccaie ....**Krishna**

"Yoga is a state of harmony and equanimity of mind"

Yogah karmasukausalam ....**Krishna**

"Yoga is observing peace . . . in the midst of activity!"

Two globally acclaimed Indian saints also defined Yoga as follows:

> "Yoga elevates man from the level of Animal Man to Divine Man."
>
> **-Sri Sri Aurobindo**

> "Each soul is potentially divine! The goal of Yoga is to manifest the divinity within by controlling the internal and external nature. This you do either by work, worship, philosophy, or psychic control, one or more or all these methods, and be free. Temples, rituals, dogmas, and doctrines are but secondary details."
>
> **-Swami Vivekananda**

Sage Patanjali, who lived in India in the second century, developed an eight—step process in his treatise Patanjali Yoga Sutras. His research is all about, "what is meant by mind" And more importantly," how do we manage our mind until it becomes absolutely subtle, tamed and is completely under our control".

The eight steps as mentioned by sage Patanjali is given below:

Step 1: Practice of Yama:

Yama means rules that are prohibitive, and refers to thoughts and actions from which the practioner should abstain. The Don'ts are injury or harming others, untruthfulness, stealing, gift receiving, noncontinence.

We need to live with the attitude that our thoughts, words or deeds should cause no harm to any other being. Our words, thoughts and action must be factual and synchronized. We need to be careful not to hurt others by saying what happens to be true but is cruel. On such occasions we have to remain silent.

In a positive sense, this means we need to cultivate the attitude of servant leadership and be ever ready to be useful to society and the less privileged beings. This is the "rent" we pay for the space we occupy in this universe—our good deed and our selfless acts.

It is not enough to simply abstain from theft; we must not harbor any feelings of covetousness, towards persons or objects. We must remember that nothing in this world belongs to us. At best we are caretakers or borrowers. It is our duty, therefore, to borrow no more from the world than we absolutely need. We need to make full and proper use of what we have. Taking more than what we need, and wasting it, is a form of stealing from the rest of mankind.

Noncontinence is nonchastity in word, thought and deed. Attachment to sex is an obstacle to progress and we need to free ourselves from this habit by cultivating the attitude of detachment.

Abstinence from greed has also been interpreted as abstinence from receiving gifts. One loses the independence of the mind on receiving gifts. In everyday world, most gifts can be regarded as relatively harmless, as long as they are tokens of genuine affection. Nevertheless, there are some gifts which are not, especially, when they belong to that rather sinister category described by income tax specialists as business gifts. We must beware, in general, of a too easy acceptance of other people's generosity and hospitality.

Step 2: Practice of Niyama

Niyama or observances are purity, contentment, Self discipline and self study (contemplation), practice and devotion to excellence.

Purity includes both physical and mental cleanliness. Physical cleanliness is important by taking a bath. Internal organs of the body must also be cleansed and strengthened by exercises, by following a proper diet and by fasting on fruit juices once a month.

We must also follow a mental diet in order in order to cleanse and strengthen the mind. We must regulate our reading, our conversation, and indeed, our whole intake of mental food. What really matters is our **"attitude?"**

The following thoughts on Attitude by Charles Swindoll should be an eye opener," The longer I live, the more I realize the impact of attitude on life. Attitude, to me, is more important than facts. It is more important than the past, than education, than money, than circumstances, than failure, than successes, than what other people think or say or do. It will make or break a company . . . . a church . . . . a home. The remarkable thing is we have choices every day regarding the attitude we will embrace for that day. We cannot change our past. We cannot change the inevitable. The only thing we can do is play on the one string we have, and that is our attitude. I am convinced that life is 10% what happens to me and 90% how I react to it. And so it is with you. We are in charge of our attitude."

If we constantly engage in exercising discrimination, we shall find that every human encounter, everything that we read or are told, has something to teach us. But this discriminative awareness is very hard to maintain, and so the beginner has to practice discrimination in his transactions every day

Every human being is searching, however confusedly, for meaning in life and hence will welcome discussion of that meaning, provided we can find a vocabulary which speaks to his or her condition. If we approach conversation from this angle and conduct it with clarity, frankness, and sincerity one develops good interpersonal skills.

Contentment means contented acceptance of one's lot in life, untroubled by envy and restlessness. Contentment does not mean passive acceptance of an unjust status quo. This would be callous indifference. Our efforts should be towards better and fairer living conditions, not only for ourselves but also for our less fortunate near and dear ones, and our neighbors. Our effort in this direction will be much more effective if it is not inspired by motives of personal gain and advantage. Contentment is about not feeling the guilt of the past, stress of the present or the worry about the future. If we observe, the days when we are totally stress free and happy, those are the days that make us wise and contented.

In life one makes progress when one makes oneself answerable to someone. In our organization, we achieve success when we make ourselves answerable to our boss, superiors and customers. We make progress in personal life when we make our self answerable to our parents; spouse or children. In fact one achieves the pinnacle of success when we make ourselves answerable to our higher self. We tend to become more disciplined. Our efforts on self study and devotion to excellence enable us to be focused and become leaders in our chosen field. Yama and Niyama yields self-control in our thoughts, words, and deeds. These practices develop in us mental calmness and a successful personality.

Step 3: Practice of Asana

Most people hold themselves badly either while walking, sitting or standing and hence are subject to all sorts of physical problems and restlessness. A good natural posture is very rarely seen in all these activities (walking, sitting, and standing). Practice of Asana enables us to take a comfortable position wherein one can sit still and erect, holding the chest, neck and head in a straight line, without strain. Energy can then be channeled to the task at hand. The aim of Asana is to achieve an effortless alertness of mind, when the body is held perfectly healthy, and steady. When body is relaxed, breath is relaxed and hence mind is at peace. The Indian spiritual scientist had discovered the truth, that when body is relaxed the mind gets automatically relaxed. Thus with the help of one's body we attain control of our mind. Practice of Asana yields both mental and physical calmness.

Step 4: Practice of Pranayama

Pranayama is not breath control as many think, but is life force control. We are continuously drawn to our external stimuli through our five senses—sight (eyes), sound (ears), smell (nose), taste (mouth) and touch sensation (skin).These stimuli trigger a lot of restlessness, and consume a lot of energy. By the practice of scientific Pranayama technique, one is consciously able to control and switch off life-force from the "five sense telephones"

The state of sleep is unconscious pranayama, or the unconscious process of switching off the life force from the "five sense telephones." In sleep, the body is still, the muscles have stopped working, venous blood is reduced, and the heart slows down, the energy in the heart becomes quiet; therefore, the energy from the five senses is switched off. One gets refreshed after a deep sleep. Patanjali does not beat around the bush but tells us to adopt the scientific pranyama technique by which sensation can be consciously disconnected from the mind. What is attained during sleep can be attained by adopting the scientific methods of pranayama. Pranayama yields heartbeat control, pulse control, awareness of cool and warm life currents in the spine, life-force control and finally, mind control.

A simple harmless breathing technique which can be used to calm the mind is right and left nostril breathing, known as Nadi Shuddi Pranayama. It can be practiced daily for 2 minutes or 10 rounds. The breathing technique is as follows. Start by exhaling through the left nostril. Inhale through the left nostril and exhale through the right nostril. Again inhale through the right nostril and exhale through the left nostril. This completes one round. Repeat this for 10 rounds. Feel the deep and slow motion of an energizing feeling throughout the whole body. With every inhalation, drive out the impurities in the body. With every exhalation enjoy the deep internal awareness, and the calm state of the mind. Allow the mind to go to deep inner state of bliss, peace and calmness.

This simple technique can do no injury, since it does not involve holding the breath excessively or over stimulating the body with too much oxygen.

The benefit of this practice is many folds: We slow down our restless mind and develop equanimity. Swami Vivekananda claims practice of this technique daily can lead to liberation of the mind.

Step 5: Practice of Pratyahara

Patanjali states that the purpose of the above four steps (Yama, Niyama, Asana, Pranayam) is to attain Pratyahara. One needs to develop the will power to control the flow of attention from the sense organs towards the sense objects. Will power can then be used to interiorize the mind on the present activity. Thousands of students are satisfied with Yama and Niyama; many are satisfied with Asana; and some are satisfied with practicing Pranayama alone. Best results come from the practice of all four namely-Yama, Niyama, Asana and Pranayama. Those who are able to quickly interiorize their mind and are undisturbed by sensations (sight, sound, smell, taste, and touch) and restless thoughts have arrived at this fifth step of the ladder i.e. Pratyahara.

Step 6: Practice of Dharana

Focusing or concentrating one's mind on one object is known as Dharana. The object of concentration can be within the body or on a picture or beautiful scenery. We normally experience this type of focus using one sense (sight) when we are deeply engrossed in reading some interesting novel or busy in some work. During such moments we are oblivious of our surrounding and totally engrossed. The mind is controlled from running in space and time and is able to focus on any object at will. With practice of Dhyana one attains mastery in focusing the mind for longer periods on one object or task.

Step 7: Practice of Dhyana

Only those who have mastered Pratyahara and Dharana can make Dhyana possible. Dhyana is all about watching the flow of thoughts. Nothing is truth—everything is exaggerated overblown thought. Even thought is manifestation of consciousness. There is a misnomer that truth is considered as a phenomenon, happening, perception, but the reality is—truth is existence, untainted by past or future events.

Countless thoughts arise in the mind like the waves in an ocean. Some thoughts absorb us while others are not of much consequence. Each thought will further carry the impulse to act. Thoughts related to parents, children, family, spouse, business, country, emotions . . . . Therefore it is essential to control and check unnecessary thoughts from occupying our mind. Unwanted thoughts can drain energy and increase the stress in an individual. Thoughts of **"Iness"** and **"Myness"** bind us to human existence.

There is a big difference between brooding over thoughts and watching over our thoughts. We brood because we identify with the thoughts and at that moment forget present reality. Watching is possible only when we observe our thoughts without attachment.

One needs to develop the art of watching our thoughts in a disengaged way, as a mute witness—every day for ten minutes—both in the morning and evening hours.

This practice is known as Dhyana. But care should be seen that backbone and head is held straight. Also breathe smoothly and be fully awake. One has a tendency to fall asleep and lose awareness during this Dhyana practice.

Sage Patanjali has elaborated in his text, Patanjali Yoga Sutras about the supernatural powers developed by the ardent practitioner of Dhyana.

Step 8: Practice of Samadhi

When the practitioner is said to be one with the object of concentration he is in the state of Samadhi. It is said in Ayurveda out of our five senses only one sense is active at one time; also they alternate so fast that one cannot make out which one sense is active. So during concentration one's thought process has to be attuned with one sense totally to achieve result. Assume you have one gallon petrol and you distribute it in 5 cars, it is doubtful whether one can move even one car a few kilometers. Similar is the case with our body, it is programmed to achieve what it wants, provideded it is only one thing.

The Samadhi state cannot be explained it has to be experienced.

The story of the devoted disciple and the deer:

On the side of a mountain in South India, overlooking a beautiful flower decked valley was a cozy hermitage. This hermitage was actually a cave, carved out of a rocky ledge of the mountain. Here dwelt a great master and a devoted disciple.

In the wee hours of the morning while dawn still lingered over the valley, the master would ask the disciple to sit upright in the perfect posture and listen to his teachings with absorbed attention. Every day the disciple eagerly devoured the lessons falling from the lips of the master. One day the King came to meet the Guru and as a parting gift handed over his pet deer. The deer was the darling of the ashram and the playful companion of the disciple. One day, however, the master noticed that his young disciple was absent minded and restless. So he gently said to him, "Son, today your mind is not on my words, it seems to be wandering over the hills. Pray tell me, what is it that causes your absentmindedness?" The disciple respectfully replied, "Honored Master, I cannot concentrate on your lessons, for my mind is helplessly thinking about our deer which is grazing on the valley."

The guru, instead of scolding the disciple, calmly asked him to retire into the chamber of silence, close the door, and think of nothing but the deer. One day passed, and the next morning the master looked through the little window in the silence chamber. The disciple was still concentrating upon the deer. So the master asked, "Son, what are you doing?" The disciple answered: "Sir I am grazing with the deer, shall I come to you." The master replied: "No, son, not yet, go on grazing with your deer."

On the third morning the master again looked through the window of the silence chamber and inquired, "Beloved child, what are you doing?" To which the disciple, in the state of ecstasy, replied: "Heavenly Master, I behold the deer in my room, and I am feeding it. Shall I come to you with the deer?" "Not yet, my son, go on with the vision of the deer, and of feeding it."

On the fifth day the master questioned "Tell me, my son, what you are doing now?" The disciple bellowed, in a deer like voice: "What do you mean? I am not your son. I am your deer." To this the master smilingly retorted: "Mr. Deer, you had better come out of the room." The disciple would not come out. "How can I get through the narrow door?" he rumbled: "My horns are too big "The master went into the chamber of silence and brought out the erstwhile "deer" out of his trance. The disciple smiled to find himself walking on all fours, trying to imitate the object of his concentration.

Then the disciple, after a light repast, went to listen to the words of his guru. He was asked many deep spiritual questions, all of which he answered correctly, as never before. At last the guru remarked: "Now your training has come to an end. Your concentration has reached the perfect state, wherein you and your mind can be one with the object of study. Use this focus in achieving success, whether in business, in the design of new products or in search for God."

Pause, Reflect, and Scribble:

____

____

____

____

____

____

____

____

____

B) Checklist for regularity and sincerity of practice.

Few guidelines

____

Review the plotted graph of peak and valley of life moments in chapter 1 and reflect with renewed understanding and insights gained;

Life is fair and gives us not what we desire but what we deserve.

The inherent nature of life is to give both pleasure and pain. That life is fair with everyone is a fact we do not realize. But humans start weeping over painful situations and start rejoicing over the pleasures. As humans we seldom feel grateful for the blessings in life. We often perceive them as our rightful due. To reduce the pain of living, we need to develop a state of perfect mental balance by which one faces all situations in life, welcoming both pleasure and pain while abiding in peace and going beyond duality considering the experiences we face as the very nature of life.

We need to implement ideas into actionable plans in order to bring about changes in our life. Mere reading and discussing or dreaming about various possibilities does not help much. Since our early childhood, we practice writing and doing most of our activities with the right hand. Suppose there is an accident and the right hand is injured, we start doing all the work with the left hand. In course of time working with our left hand becomes natural to us. A musician starts tapping his hand the moment he hears music. The power of practice is so powerful, in fact even reason is considered subordinate to the power of practice. Through practice we can develop habits that stay with us even in times of great stress or danger whereas reason can simply desert us at such times. This is the reason doctors and nurses can save people, while others panic at the sight of blood. Doctors and nurses are made to practice the procedures so many times they are able to perform involuntarily on the accident patients and save them from death. So effort and practice of learning depends on our sense of urgency, necessity and utility.

Think of the amazing feats performed by circus artists, riding a horse standing on one leg, walking on a thin rope that hangs sixty feet above the ground, and so on. How are they able to perform these feats without fear? Practicing the same thing over and over again for years has made them perfect. Practice makes a man perfect.

Genius is 1 percent inspiration and 99 percent perspiration. If you had seen the famous scientist during their childhood days struggling to learn alphabets, would you have ever thought that they would become so famous? Did Einstein ever think that he would become so great when he was still a small child, learning simple lessons in the class? It is only aspiration and working on their purpose that transformed them into great people.

We can achieve the impossible and achieve greatness through practice, patience and perseverance. Lack of proper resolve and Lack of strong will power stand in the way of achieving greatness. We are not able to improve our lives because we lack earnestness in putting the right efforts, and a lack of interest in keeping up the practice. Most people start off with good intentions but after a few days practice, they get bored or disheartened and give up. That is why one fails in so many of one's endeavors. Age—old wisdom says that if one takes up practicing something noble, it should not be given up even for a day. The danger is that if we neglect our practice even once, the mind will invent some excuse or the other to skip the practice in future as well. Our mind must not be given any lenience. One must have enough patience and perseverance to keep up the practice steadily for a long time.

Remember to practice Swami Vivekananda's statement, "One ounce of practice is worth a thousand pounds of theory. Practice makes us what we shall be."

All benefits come without fail through regular practice. Avoid blaming other people or circumstance for problems one faces. Strive to recognize that we ourselves are the creator of every situation in our life and destiny. It appears to me that almost any man—like the spider spins from its own volition—can spin his success and his destiny. When you recognize your shortcomings, do not be discouraged or vexed over it; rather be determined to overcome it with Spirit of Inquiry and Spirit of Awareness.

Given below few must do checklists:

- **Look forward to doing ten routine things everyday with passion and energy.**
- **Try learning about one new thing every week.**

- **Work on your purpose for thousand days for reaping the benefits.**
- **Practice three dimensional awareness: namely physical awareness, mental awareness and social awareness for unfolding your potential.**
- **Put your entire attention in enjoying what you are doing (Now).**
- **Practice treating people as people not as objects.**
- **Practice giving more than what you receive.**
- **Connect with others, expand your consciousness, and see yourself in others and others in you.**
- **Be with the flow of life.**
- **Lower your defenses in transactions and the concept of who you are.**
- **Practice equanimity and the art of absorptive listening.**
- **Remind yourself you are breathless and ageless.**
- **Practice! Do it now! No Procrastination.**
- **Stop taking for granted, life's gift that comes free. Pay with gratitude, service and contentment.**
- **Smile for changing your mood; you make a difference to the world.**

Keep on practicing the above must do checklist. Even if you fall short of your expectation, continuous effort regardless of temporary failure will ultimately carry the faithful individual to the Ultimate Goal;

- **Balanced Living**
- **Achieving your purpose**
- **Unfolding your potential**
- **Realizing your wildest dreams . . . . !**

Summary

Strive for Balanced Living

Most of us spend our lives as if we had another one in the bank.

-Ben Irwin

- Our true nature is pure knowledge, pure consciousness and pure bliss . . . . Satchitananda.
- Each Life—spark is potentially divine, hence respect life in all forms.
- Excess speed and restlessness brings imbalance and leads to deterioration in the quality of life and impairs efficiency.
- For balanced living, practice silence, practice nonjudgmental, practice non—résistance, practice yoga.
- Yoga sheds light on the body, mind complexity and hence is a solution for all our physical and psychological problems.
- Yoga messages or understanding cannot be taught like history or mathematics; it is transmitted like light or heat.
- Sage Patanjali has shared eight—step process namely Yama, Niyama, Asana, Pranayama, Pratyahara, Dharna, Dhyana, Samadhi for controlling and taming our mind.

References for further Reading:

Sri Paramahamsa Yogananda-God talks to Arjuna,

Deepak Chopra-Seven laws of success

Patanjali Yoga Sutras.

COMMENTS FROM READERS

The book though small is a true expression of Professor Lakshman. In this book, he has bought out some of the key essence and key aspects of a leader. Instead of delving on the styles of leadership, the book goes deeper and explores the common denominators of true leaders—self inquiry, purpose, breaking free from habits and so on. There is a strange sense of compassion and a sense of non judgment in the book. The book also is very personal and every reader can personalize and create their own space in this book.

Without getting pedantic and preaching, it has a smooth flow, with the right number of case studies, stories and simple messages. What I liked most about the book is its honesty, simplicity and a true desire to communicate oneself to the world.

A young reader, a college student, housewives and individuals from all walks of life will find something in this book that touches their heart. For all the readers—I would recommend just one thing—read it from the heart, connect to the soul of the book and maybe you will feel the depth within.

Ms Rashmi Aiyappa,

Spiritual Scientist-Ms Rashmi Aiyappa is the founder of a movement called Aashwasan. She is globally renowned spiritual scientist, and has helped transform lives of thousands of people all over the world."

This book, by Prof Lakshman, `Life is Balanced Living is not!' Creates the spirit of self-awareness in individuals which will help mould a higher destiny for themselves. When self-awareness is lacking, people will continue to wallow in the same loop of `garbage in garbage out'. I appreciate Prof Lakshman's deep interest in imparting the spirit of inquiry and spirit of awareness among students and people of all ages

A Senthivel-Author, Life is Fundamentally Management.

This book, Life is balanced Living is Not! makes interesting reading. I love the scribble area. A lot of anecdotes & short narrates. I would definitely buy the book and enjoy reading.

Geeta Ramakrishnan-Executive Director, Transworld Shipping.

I enjoyed reading the book-lots of good advice-will need a lot of discipline to put it to practice.

Adithya Sharma—Student-University of Washington DC

Overall the book, Life is Balanced Living is Not! is well written and could be used and adapted as a course material for colleges. There are lots of intense chapters like chapter #8 which needs some knowledge &understanding of Indian culture.

Just reading the book is not sufficient. One needs to reflect & practice the concepts mentioned by Prof Lakshman for the transformation and magic to happen.

Venkat krisna-Project Manager-Porsche US.

In a simple, beautiful and artistic manner the book, Life is Balanced Living is Not! is crafted successfully. The contents are impressive and covers the art of balancing life at emotional, intellectual and awareness level. Prof Lakshman provides the reader with a practical guide for balanced Living.

Yours in Dhamma
Dhammapala Guruji-Maha Bodhi Society.

I had the pleasure of reading the contents of the book, Life is Balanced Living is Not! authored by Prof Lakshman. The coherence and lucidity with which the content has been developed with apt examples is a salient feature of this book.

I am also impressed the way in which the principles of success and leadership have been identified by the author.

This book is worth reading and may be adopted as a course," **Principles of Self Science"** for graduate and post graduate programme.

Dr N.Sundararajan,
Vice Chancellor,Jain University.

A well knit book that can serve as a self –help guide for personal growth.

The stories convey messages that are appropriate and thought provoking.

Prof NVH Krishnan,
Registrar Jain University
Director,Arka Eduservice Ltd